NASA/TP—1999–209484

Total Solar Eclipse of 2001 June 21

Fred Espenak and Jay Anderson

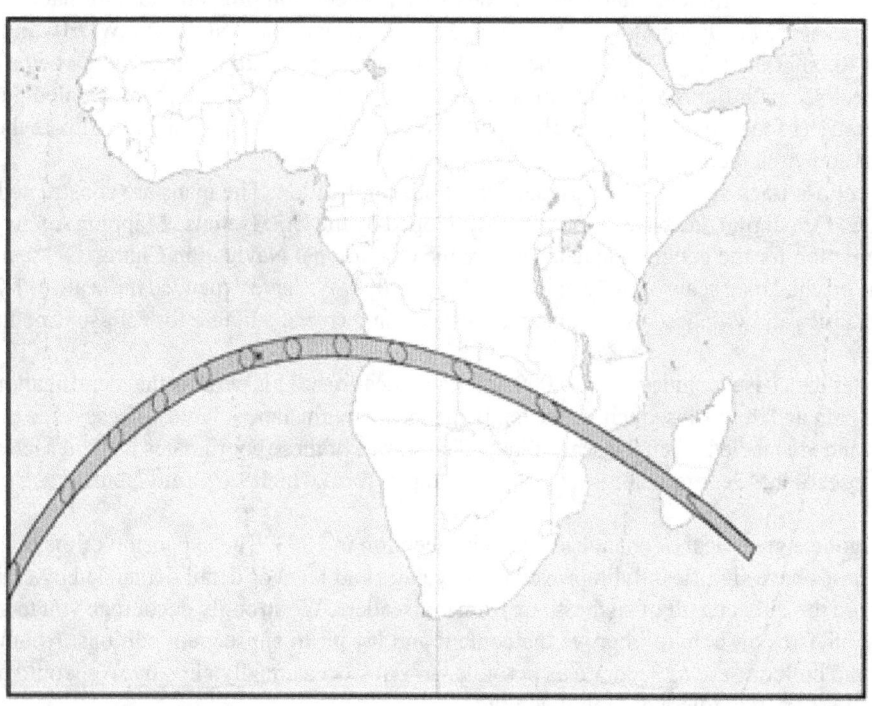

National Aeronautics and
Space Administration

Goddard Space Flight Center
Greenbelt, Maryland 20771

November 1999

PREFACE

This work is the latest in a series of NASA publications containing detailed predictions, maps and meteorological data for future central solar eclipses of interest. Published as part of NASA's Technical Publication (TP) series[1], the eclipse bulletins are prepared in cooperation with the Working Group on Eclipses of the International Astronomical Union and are provided as a public service to both the professional and lay communities, including educators and the media. In order to allow a reasonable lead time for planning purposes, eclipse bulletins are published 18 to 24 months before each event.

Single copies of the bulletins are available at no cost and may be ordered by sending a 9 x 12 inch self addressed stamped envelope (SASE) with sufficient postage (12 oz. or 340 g.). Use stamps only; cash or checks cannot be accepted. Requests within the U. S. may use the Postal Service's Priority Mail for $3.20. Please print the eclipse date (year & month) or the NASA publication number in the lower left corner of the return SASE. Requests from outside the U. S. and Canada may send ten international postal coupons to cover postage. Exceptions to the postage requirements will be made to professional researchers and scientists, or for international requests where political or economic restraints prevent the transfer of funds to other countries. All requests should be accompanied by a copy of the request form on the last page.

The 2001 bulletin uses two mapping data bases for the path figures: World Vector Shoreline (WVS) and World Data Bank II (WDBII). The WDBII outline files were digitized from navigational charts to a scale of approximately 1:3,000,000, and represent the "state of the art" in the 1970s. The WVS data are given at several resolutions, including 1:1,000,000, 1:250,000 and 1:100,000. For maximum efficiency and speed, these data have been compressed and reformatted into direct access files by Jan C. Depner (U. S. Naval Oceanographic Office) and James A. Hammak (NORDA). WDBII and WVS are available through the Global Relief Data CD-ROM from the National Geophysical Data Center. These vector data have made it possible to generate eclipse path figures at resolutions greater than 1:10,000,000. The more detailed path figures include curves of constant duration of totality within the umbral path. This permits the user to quickly estimate the duration of totality at various locations shown in the figures.

High detail maps of the eclipse track appear in the last section of this publication. The maps are constructed from the Digital Chart of the World (DCW), a digital database of the world developed by the U.S. Defense Mapping Agency (DMA). The primary sources of information for the geographic database are the Operational Navigation Charts (ONC) and the Jet Navigation Charts (JNC). The original map scale of these maps was 1:1,000,000. We have expanded the scale to 1:2,500,000 in order to show major roads, cities and villages, coarse topography, lakes and rivers, suitable for eclipse expedition planning.

The geographic coordinates data base includes over 90,000 cities and locations. This permits the identification of many more cities within the umbral path and their subsequent inclusion in the local circumstances tables. These same coordinates are plotted in the path figures and are labeled when the scale allows. The source of these coordinates is Rand McNally's The New International Atlas. A subset of these coordinates is available in a digital form which we've augmented with population data.

The bulletins have undergone a great deal of change since their inception in 1993. The expansion of the mapping and geographic coordinates data bases have significantly improved the coverage and level of detail demanded by eclipse planning. Some of these changes are the direct result of suggestions from our readers. We strongly encourage you to share your comments, suggestions and criticisms on how to improve the content and layout in subsequent editions. Although every effort is made to ensure that the bulletins are as accurate as possible, an error occasionally slips by. We would appreciate your assistance in reporting all errors, regardless of their magnitude.

We thank Dr. B. Ralph Chou for a comprehensive discussion on solar eclipse eye safety. Dr. Chou is Professor of Optometry at the University of Waterloo and he has over twenty-five years of eclipse observing experience. As a leading authority on the subject, Dr. Chou's contribution should help dispel much of the fear and misinformation about safe eclipse viewing.

Dr. Joe Gurman (GSFC/Solar Physics Branch) has made this and previous eclipse bulletins available over the Internet. They can be read or downloaded via the World-Wide Web from Goddard's Solar Data Analysis Center eclipse information page: http://umbra.nascom.nasa.gov/eclipse/.

During 1996, Espenak developed a new web site which provides general information on both solar and lunar eclipses occurring during the next two decades. Hints on eclipse photography and eye safety may be found there as well as links to other eclipse related web sites. The URL for the site is: http://sunearth.gsfc.nasa.gov/eclipse/eclipse.html.

In addition to the general information web site above, a special web site devoted to the 2001 total solar eclipse has been set up: http://sunearth.gsfc.nasa.gov/eclipse/TSE2001/TSE2001.html. It includes supplemental predictions, figures and maps which could not be included in the present publication.

---

[1]Previous bulletins were published as NASA Reference Publications. That series has now been replaced by the NASA Technical Publication series.

Since the eclipse bulletins are of a limited and finite size, they cannot include everything needed by every scientific investigation. Some investigators may require exact contact times which include lunar limb effects or for a specific observing site not listed in the bulletin. Other investigations may need customized predictions for an aerial rendezvous or from the path limits for grazing eclipse experiments. We would like to assist such investigations by offering to calculate additional predictions for any professionals or large groups of amateurs. Please contact Espenak with complete details and eclipse prediction requirements.

We would like to acknowledge the valued contributions of a number of individuals who were essential to the success of this publication. The format and content of the NASA eclipse bulletins has drawn heavily upon over 40 years of eclipse Circulars published by the U. S. Naval Observatory. We owe a debt of gratitude to past and present staff of that institution who have performed this service for so many years. The many publications and algorithms of Dr. Jean Meeus have served to inspire a life-long interest in eclipse prediction. We thank Francis Reddy, who helped develop the original data base of geographic coordinates and to Rique Pottenger (Astro Communications Service) for his assistance in expanding the data base to over 90,000 cities. Internet availability of the eclipse bulletins is due to the efforts of Dr. Joseph B. Gurman. The support of Environment Canada is acknowledged in the acquisition of the weather data. Permission is freely granted to reproduce any portion of this publication, including data, figures, maps, tables and text.

All uses and/or publication of this material should be accompanied by an appropriate acknowledgment (e.g., - "Reprinted from Total Solar Eclipse of 2001 June 21, Espenak and Anderson, 1999"). We would appreciate receiving a copy of any publications where this material appears.

The names and spellings of countries, cities and other geopolitical regions are not authoritative, nor do they imply any official recognition in status. Corrections to names, geographic coordinates and elevations are actively solicited in order to update the data base for future eclipses. All calculations, diagrams and opinions are those of the authors and they assume full responsibility for their accuracy.

Fred Espenak
NASA/Goddard Space Flight Center
Planetary Systems Branch, Code 693
Greenbelt, MD 20771
USA

email: espenak@gsfc.nasa.gov
FAX: (301) 286-0212

Jay Anderson
Environment Canada
123 Main Street, Suite 150
Winnipeg, MB,
CANADA R3C 4W2

email: jander@cc.umanitoba.ca
FAX: (204) 983-0109

Current and Future NASA Solar Eclipse Bulletins

NASA Eclipse Bulletin	RP #	Publication Date
Annular Solar Eclipse of 1994 May 10	*1301*	*April 1993*
Total Solar Eclipse of 1994 November 3	*1318*	*October 1993*
Total Solar Eclipse of 1995 October 24	*1344*	*July 1994*
Total Solar Eclipse of 1997 March 9	*1369*	*July 1995*
Total Solar Eclipse of 1998 February 26	*1383*	*April 1996*
Total Solar Eclipse of 1999 August 11	*1398*	*March 1997*
Total Solar Eclipse of 2001 June 21	*1999-209484*	*November 1999*

- - - - - - - - - - - future - - - - - - - - - - -

Total Solar Eclipse of 2002 December 4	*—*	*late 2000*
Annular and Total Solar Eclipses of 2003	*—*	*late 2001*
Transit of Venus of 2004 June 8	*—*	*2002*

TOTAL SOLAR ECLIPSE OF 2001 JUNE 21

Table of Contents

TOTAL SOLAR ECLIPSE OF 2001 JUNE 21

Figures, Tables and Maps

Figures, Tables and Maps - CONTINUED

ECLIPSE PREDICTIONS

INTRODUCTION

On Thursday, 2001 June 21, a total eclipse of the Sun will be visible from within a narrow corridor which traverses the Southern Hemisphere. The path of the Moon's umbral shadow begins in the South Atlantic, crosses southern Africa and Madagascar, and ends at sunset in the Indian Ocean. A partial eclipse will be seen within the much broader path of the Moon's penumbral shadow, which includes eastern South America and the southern two thirds of Africa (Figures 1 and 2).

UMBRAL PATH AND VISIBILITY

The first total solar eclipse of the Third Millennium begins in the South Atlantic about 400 kilometers southeast of Uruguay . The Moon's umbral shadow first touches down on Earth at 10:35:55 UT. Along the sunrise terminator, the duration is 2 minutes 6 seconds as seen from the center of the 127 kilometers wide path. During the next two hours, the umbra sweeps across the South Atlantic with no major landfall. The central duration, path width and Sun's altitude all steadily increase as the Moon's shadow rushes along its northeastern transoceanic course.

The instant of greatest eclipse[1] occurs at 12:03:41 UT when the axis of the Moon's shadow passes closest to the center of Earth (gamma[2] = -0.570). The length of totality reaches its maximum duration of 4 minutes 56 seconds, the Sun's altitude is 55°, the path width is 200 kilometers and the umbra's velocity is 0.554 km/s. Unfortunately, these most favorable circumstances occur at sea some 1100 kilometers west of equatorial Africa's South Atlantic coastline (Figure 3).

Another half hour will elapse before the umbra reaches southern Africa. During this period, the duration and path width will slowly wane as the curvature of Earth carries the projected eclipse path away from the Moon. As a result, the tapering shadow cone will present a progressively narrower cross section to Earth's surface along the path of totality. Nevertheless, the 2001 total solar eclipse is over twice as long as the last total eclipse in 1999 at all corresponding positions along the path (i.e. - sunrise, transit, sunset).

Landfall finally occurs along the Atlantic coast of Angola at 12:36 UT (Figure 4). The center line duration is 4 minutes 36 seconds, the path width is 193 kilometers and the Sun's altitude is 49°. Traveling with a ground speed of 0.63 km/s, the umbra swiftly moves eastward through the civil war torn nation. Although the weather prospects are excellent, Angola's political strife makes it a dangerous place for eclipse observers. By 12:57 UT, the shadow reaches the western border of Zambia where totality lasts a maximum of 4 minutes 6 seconds (Figure 5). The umbra's path is now directed to the east-southeast as it heads towards the nation's capitol city. Lying about 35 km south of the center line, Lusaka enjoys a total eclipse lasting a respectable 3 minutes 14 seconds. Further to the north, the center line duration is 3 minutes 35 seconds and the Sun's altitude is 31°. Quickly crossing the Zambezi River, the path continues into Zimbabwe and Mozambique. The center line runs near the border between these two nations for several hundred kilometers. Zimbabwe's capitol city of Harare lies just 100 km south of the path, so its citizens will witness a partial eclipse of magnitude 0.976 (Figure 6). As the umbra leaves Zambia, it lies wholly within Mozambique for a minute before its leading edge reaches the shore of the Indian Ocean (~13:20 UT). At nearly the same time, its northern edge just skirts the southern border of Malawi. The center line duration is now 3 minutes 09 seconds as the Sun's altitude drops to 23°. The shadow's speed has increased to 1.7 km/s as it begins to leave the continent for the open waters of the Indian Ocean.

One last major landfall remains as the umbra races over southern Madagascar (Figure 7). The eclipse occurs quite late in the day with the Sun just 11° above the western horizon. The central duration now stands at 2 minutes 25 seconds (13:28 UT) as the shadow's long elliptical projection stretches across the entire east-west breadth of the island. Continuing into the Indian Ocean, the umbra leaves Earth's surface (13:31:33 UT) as the path ends. Over the course of 2 hours and 54 minutes, the Moon's umbra travels along a path approximately 12,000 kilometers long and covering 0.3% of Earth's surface area.

---

[1] The instant of greatest eclipse occurs when the distance between the Moon's shadow axis and Earth's geocenter reaches a minimum. Although greatest eclipse differs slightly from the instants of greatest magnitude and greatest duration (for total eclipses), the differences are usually quite small.

[2] Minimum distance of the Moon's shadow axis from Earth's center in units of equatorial Earth radii.

GENERAL MAPS OF THE ECLIPSE PATH

ORTHOGRAPHIC PROJECTION MAP OF THE ECLIPSE PATH

Figure 1 is an orthographic projection map of Earth [adapted from Espenak, 1987] showing the path of penumbral (partial) and umbral (total) eclipse. The daylight terminator is plotted for the instant of greatest eclipse with north at the top. The sub-Earth point is centered over the point of greatest eclipse and is indicated with an asterisk-like symbol. The sub-solar point (Sun in zenith) at that instant is also shown.

The limits of the Moon's penumbral shadow define the region of visibility of the partial eclipse. This saddle shaped region often covers more than half of Earth's daylight hemisphere and consists of several distinct zones or limits. At the northern and/or southern boundaries lie the limits of the penumbra's path. Partial eclipses have only one of these limits, as do central eclipses when the shadow axis falls no closer than about 0.45 radii from Earth's center. Great loops at the western and eastern extremes of the penumbra's path identify the areas where the eclipse begins/ends at sunrise and sunset, respectively. If the penumbra has both a northern and southern limit, the rising and setting curves form two separate, closed loops. Otherwise, the curves are connected in a distorted figure eight. Bisecting the 'eclipse begins/ends at sunrise and sunset' loops is the curve of maximum eclipse at sunrise (western loop) and sunset (eastern loop). The exterior tangency points **P1** and **P4** mark the coordinates where the penumbral shadow first contacts (partial eclipse begins) and last contacts (partial eclipse ends) Earth's surface. The path of the umbral shadow bisects the penumbral path from west to east and is shaded dark gray.

A curve of maximum eclipse is the locus of all points where the eclipse is at maximum at a given time. They are plotted at each half hour Universal Time (UT), and generally run from northern to southern penumbral limits, or from the maximum eclipse at sunrise or sunset curves to one of the limits. The outline of the umbral shadow is plotted every ten minutes in UT. Curves of constant eclipse magnitude[3] delineate the locus of all points where the magnitude at maximum eclipse is constant. These curves run exclusively between the curves of maximum eclipse at sunrise and sunset. Furthermore, they are quasi-parallel to the northern/southern penumbral limits and the umbral paths of central eclipses. Northern and southern limits of the penumbra may be thought of as curves of constant magnitude of 0%, while adjacent curves are for magnitudes of 20%, 40%, 60% and 80%. The northern and southern limits of the path of total eclipse are curves of constant magnitude of 100%.

At the top of Figure 1, the Universal Time of geocentric conjunction between the Moon and Sun is given followed by the instant of greatest eclipse. The eclipse magnitude is given for greatest eclipse. For central eclipses (both total and annular), it is equivalent to the geocentric ratio of diameters of the Moon and Sun. Gamma is the minimum distance of the Moon's shadow axis from Earth's center in units of equatorial Earth radii. The shadow axis passes south of Earth's geocenter for negative values of Gamma. Finally, the Saros series number of the eclipse is given along with its relative sequence in the series.

STEREOGRAPHIC PROJECTION MAP OF THE ECLIPSE PATH

The stereographic projection of Earth in Figure 2 depicts the path of penumbral and umbral eclipse in greater detail. The map is oriented with the north up with the point of greatest eclipse near the center. International political borders are shown and circles of latitude and longitude are plotted at 20° increments. The region of penumbral or partial eclipse is identified by its northern and southern limits, curves of eclipse begins or ends at sunrise and sunset, and curves of maximum eclipse at sunrise and sunset. Curves of constant eclipse magnitude are plotted for 20%, 40%, 60% and 80%, as are the limits of the path of total eclipse. Also included are curves of greatest eclipse at every half hour Universal Time.

Figures 1 and 2 may be used to determine quickly the approximate time and magnitude of maximum eclipse at any location within the eclipse path.

EQUIDISTANT CONIC PROJECTION MAP OF THE ECLIPSE PATH

Figure 3 is an equidistant conic projection map chosen to minimize distortion, and which isolates a specific region of the umbral path. Once again, curves of maximum eclipse and constant eclipse magnitude are plotted and labeled. A linear scale is included for estimating approximate distances

---

[3] Eclipse magnitude is defined as the fraction of the Sun's diameter occulted by the Moon. It is strictly a ratio of *diameters* and should not be confused with eclipse obscuration which is a measure of the Sun's surface *area* occulted by the Moon. Eclipse magnitude may be expressed as either a percentage or a decimal fraction (e.g.: 50% or 0.50).

(kilometers). Within the northern and southern limits of the path of totality, the outline of the umbral shadow is plotted at ten minute intervals. The duration of totality (minutes and seconds) and the Sun's altitude correspond to the local circumstances on the center line at each shadow position.

The scale used in this figure is ~1:31,804,000. The positions of larger cities and metropolitan areas in and near the umbral path are depicted as black dots. The size of each city is logarithmically proportional to its population using 1990 census data (Rand McNally, 1991).

EQUIDISTANT CYLINDRICAL PROJECTION MAPS OF THE ECLIPSE PATH

Figures 4 through 7 all use a simple equidistant cylindrical projection scaled for the central latitude of each map. They all use high resolution coastline data from the World Data Base II (WDB) and World Vector Shoreline (WVS) data bases and have a scale of 1:4,452,000. These maps were chosen to isolate small regions along the entire land portion of the eclipse path. Once again, curves of maximum eclipse and constant eclipse magnitude are included as well as the outline of the umbral shadow. A special feature of these maps are the curves of constant umbral eclipse duration (i.e., totality) which are plotted within the path. These curves permit fast determination of approximate durations without consulting any tables. Furthermore, city data from a recently enlarged geographic data base of over 90,000 positions are plotted to give as many locations as possible in the path of totality. Local circumstances have been calculated for these positions and can be found in Tables 9 through 15.

ELEMENTS, SHADOW CONTACTS AND ECLIPSE PATH TABLES

The geocentric ephemeris for the Sun and Moon, various parameters, constants, and the Besselian elements (polynomial form) are given in Table 1. The eclipse elements and predictions were derived from the DE200 and LE200 ephemerides (solar and lunar, respectively) developed jointly by the Jet Propulsion Laboratory and the U. S. Naval Observatory for use in the *Astronomical Almanac* for 1984 and thereafter. Unless otherwise stated, all predictions are based on center of mass positions for the Moon and Sun with no corrections made for center of figure, lunar limb profile or atmospheric refraction. The predictions depart from normal IAU convention through the use of a smaller constant for the mean lunar radius k for all umbral contacts (see: LUNAR LIMB PROFILE). Times are expressed in either Terrestrial Dynamical Time (TDT) or in Universal Time (UT), where the best value of ΔT[4] available at the time of preparation is used.

From the polynomial form of the Besselian elements, any element can be evaluated for any time t_1 (in decimal hours) via the equation:

$$\mathbf{a} = \mathbf{a}_0 + \mathbf{a}_1 * t + \mathbf{a}_2 * t^2 + \mathbf{a}_3 * t^3 \quad \text{(or } \mathbf{a} = \Sigma\,[\mathbf{a}_n * t^n]; \text{ n = 0 to 3)}$$

where: $\mathbf{a} = x, y, d, l_1, l_2, \text{ or } \mu$
$t = t_1 - t_0$ (decimal hours) and $t_0 = 12.00$ TDT

The polynomial Besselian elements were derived from a least-squares fit to elements rigorously calculated at five separate times over a six hour period centered at t_0. Thus, the equation and elements are valid over the period $9.00 \le t_1 \le 15.00$ TDT.

Table 2 lists all external and internal contacts of penumbral and umbral shadows with Earth. They include TDT times and geodetic coordinates with and without corrections for ΔT. The contacts are defined:

P1 - Instant of first external tangency of penumbral shadow cone with Earth's limb.
    (partial eclipse begins)
P4 - Instant of last external tangency of penumbral shadow cone with Earth's limb.
    (partial eclipse ends)
U1 - Instant of first external tangency of umbral shadow cone with Earth's limb.
    (umbral eclipse begins)
U2 - Instant of first internal tangency of umbral shadow cone with Earth's limb.
U3 - Instant of last internal tangency of umbral shadow cone with Earth's limb.
U4 - Instant of last external tangency of umbral shadow cone with Earth's limb.
    (umbral eclipse ends)

---

[4] ΔT is the difference between Terrestrial Dynamical Time and Universal Time

Similarly, the northern and southern extremes of the penumbral and umbral paths, and extreme limits of the umbral center line are given. The IAU longitude convention is used throughout this publication (i.e., for longitude, east is positive and west is negative; for latitude, north is positive and south is negative).

The path of the umbral shadow is delineated at five minute intervals in Universal Time in Table 3. Coordinates of the northern limit, the southern limit and the center line are listed to the nearest tenth of an arc-minute (~185 m at the Equator). The Sun's altitude, path width and umbral duration are calculated for the center line position. Table 4 presents a physical ephemeris for the umbral shadow at five minute intervals in UT. The center line coordinates are followed by the topocentric ratio of the apparent diameters of the Moon and Sun, the eclipse obscuration[5], and the Sun's altitude and azimuth at that instant. The central path width, the umbral shadow's major and minor axes and its instantaneous velocity with respect to Earth's surface are included. Finally, the center line duration of the umbral phase is given.

Local circumstances for each center line position listed in Tables 3 and 4 are presented in Table 5. The first three columns give the Universal Time of maximum eclipse, the center line duration of totality and the altitude of the Sun at that instant. The following columns list each of the four eclipse contact times followed by their related contact position angles and the corresponding altitude of the Sun. The four contacts identify significant stages in the progress of the eclipse. They are defined as follows:

First Contact — Instant of first external tangency between the Moon and Sun. (partial eclipse begins)

Second Contact — Instant of first internal tangency between the Moon and Sun. (central or umbral eclipse begins; total or annular eclipse begins)

Third Contact — Instant of last internal tangency between the Moon and Sun. (central or umbral eclipse ends; total or annular eclipse ends)

Fourth Contact — Instant of last external tangency between the Moon and Sun. (partial eclipse ends)

The position angles **P** and **V** identify the point along the Sun's disk where each contact occurs[6]. Second and third contact altitudes are omitted since they are always within 1° of the altitude at maximum eclipse.

Table 6 presents topocentric values from the central path at maximum eclipse for the Moon's horizontal parallax, semi-diameter, relative angular velocity with respect to the Sun, and libration in longitude. The altitude and azimuth of the Sun are given along with the azimuth of the umbral path. The northern limit position angle identifies the point on the lunar disk defining the umbral path's northern limit. It is measured counter-clockwise from the north point of the Moon. In addition, corrections to the path limits due to the lunar limb profile are listed. The irregular profile of the Moon results in a zone of 'grazing eclipse' at each limit that is delineated by interior and exterior contacts of lunar features with the Sun's limb. This geometry is described in greater detail in the section LIMB CORRECTIONS TO THE PATH LIMITS: GRAZE ZONES. Corrections to center line durations due to the lunar limb profile are also included. When added to the durations in Tables 3, 4, 5 and 7, a slightly shorter central total phase is predicted along most of the path.

To aid and assist in the plotting of the umbral path on large scale maps, the path coordinates are also tabulated at 1° intervals in longitude in Table 7. The latitude of the northern limit, southern limit and center line for each longitude is tabulated to the nearest hundredth of an arc-minute (~18.5 m at the Equator) along with the Universal Time of maximum eclipse at each position. Finally, local circumstances on the center line at maximum eclipse are listed and include the Sun's altitude and azimuth, the umbral path width and the central duration of totality.

In applications where the zones of grazing eclipse are needed in greater detail, Table 8 lists these coordinates over land based portions of the path at 30' intervals in longitude. The time of maximum eclipse is given at both northern and southern limits as well as the path's azimuth. The elevation and scale factors are also given (See: LIMB CORRECTIONS TO THE PATH LIMITS: GRAZE ZONES).

---

[5] Eclipse obscuration is defined as the fraction of the Sun's surface area occulted by the Moon.

[6] P is defined as the contact angle measured counter-clockwise from the *north* point of the Sun's disk.

V is defined as the contact angle measured counter-clockwise from the *zenith* point of the Sun's disk.

LOCAL CIRCUMSTANCES TABLES

Local circumstances for approximately 350 cities, metropolitan areas and places in South America, and Africa are presented in Tables 9 through 15. These tables give the local circumstances at each contact and at maximum eclipse[7] for every location. The coordinates are listed along with the location's elevation (meters) above sea-level, if known. If the elevation is unknown (i.e., not in the data base), then the local circumstances for that location are calculated at sea-level. In any case, the elevation does not play a significant role in the predictions unless the location is near the umbral path limits and the Sun's altitude is relatively small (<10°). The Universal Time of each contact is given to a tenth of a second, along with position angles **P** and **V** and the altitude of the Sun. The position angles identify the point along the Sun's disk where each contact occurs and are measured counter-clockwise (i.e., eastward) from the north and zenith points, respectively. Locations outside the umbral path miss the umbral eclipse and only witness first and fourth contacts. The Universal Time of maximum eclipse (either partial or total) is listed to a tenth of a second. Next, the position angles **P** and **V** of the Moon's disk with respect to the Sun are given, followed by the altitude and azimuth of the Sun at maximum eclipse. Finally, the corresponding eclipse magnitude and obscuration are listed. For umbral eclipses (both annular and total), the eclipse magnitude is identical to the topocentric ratio of the Moon's and Sun's apparent diameters.

Two additional columns are included if the location lies within the path of the Moon's umbral shadow. The **umbral depth** is a relative measure of a location's position with respect to the center line and path limits. It is a unitless parameter which is defined as:

$$\mathbf{u} = 1 - \text{abs}(\mathbf{x}/\mathbf{R}) \tag{1}$$

where: $\mathbf{u}$ = umbral depth
$\mathbf{x}$ = perpendicular distance from the shadow axis (kilometers)
$\mathbf{R}$ = radius of the umbral shadow as it intersects Earth's surface (kilometers)

The umbral depth for a location varies from 0.0 to 1.0. A position at the path limits corresponds to a value of 0.0 while a position on the center line has a value of 1.0. The parameter can be used to quickly determine the corresponding center line duration. Thus, it is a useful tool for evaluating the trade-off in duration of a location's position relative to the center line. Using the location's duration and umbral depth, the center line duration is calculated as:

$$\mathbf{D} = \mathbf{d} / (1 - (1 - \mathbf{u})^2)^{1/2} \text{ seconds} \tag{2}$$

where: $\mathbf{D}$ = duration of totality on the center line (seconds)
$\mathbf{d}$ = duration of totality at location (seconds)
$\mathbf{u}$ = umbral depth

The final column gives the duration of totality. The effects of refraction have not been included in these calculations, nor have there been any corrections for center of figure or the lunar limb profile.

Locations were chosen based on general geographic distribution, population, and proximity to the path. The primary source for geographic coordinates is *The New International Atlas* (Rand McNally, 1991). Elevations for major cities were taken from *Climates of the World* (U. S. Dept. of Commerce, 1972). In this rapidly changing political world, it is often difficult to ascertain the correct name or spelling for a given location. Therefore, the information presented here is for location purposes only and is not meant to be authoritative. Furthermore, it does not imply recognition of status of any location by the United States Government. Corrections to names, spellings, coordinates and elevations is solicited in order to update the geographic data base for future eclipse predictions.

For countries in the path of totality, expanded versions of the local circumstances tables listing many more locations are available via a special web site of supplemental material for the total solar eclipse of 1999 (*http://sunearth.gsfc.nasa.gov/eclipse/TSE2001/TSE2001.html*).

---

[7] For partial eclipses, maximum eclipse is the instant when the greatest fraction of the Sun's diameter is occulted. For total eclipses, maximum eclipse is the instant of mid-totality.

DETAILED MAPS OF THE UMBRAL PATH

The path of totality has been plotted on a set of ten detailed maps appearing in the last section of this publication. The maps are constructed from the Digital Chart of the World (DCW), a digital database of the world developed by the U.S. Defense Mapping Agency (DMA). The primary sources of information for the geographic database are the Operational Navigation Charts (ONC) and the Jet Navigation Charts (JNC) developed by the DMA. The original map scale of these maps was 1:1,000,000. Previous users of these publications will be familiar with ONC and JNC charts as hard-copy versions were used to show the eclipse paths in past bulletins.

The scale of these maps has been increased to 1:2,500,000, adequate for showing major transportation routes, cities and villages, coarse topography, lakes and rivers, suitable for eclipse expedition planning. Caution should be employed in using these maps as no distinction has been made between major highways and second class soft-surface roads in the map plot. Those eclipse viewers who require more detailed plots of the eclipse track should use the data contained within the tables in this publication and larger scale background maps.

The DCW database was assembled in the 1980s and contains names of places that are no longer used in some parts of Africa, particularly Zimbabwe. Where possible, modern names have been substituted for those in the database but this correction could not be applied to all sites. Some areas of missing topographic data appear as blank white rectangles on the map background.

Northern and southern limits as well as the center line of the path are plotted using data from Table 7. Although no corrections have been made for center of figure or lunar limb profile, they have little or no effect at this scale. Atmospheric refraction has not been included, as its effects play a significant role only at very low solar altitudes. In any case, refraction corrections to the path are uncertain since they depend on the atmospheric temperature-pressure profile, which cannot be predicted in advance. If observations from the graze zones are planned, then the zones of grazing eclipse must be plotted on higher scale maps using coordinates in Table 8. See PLOTTING THE PATH ON MAPS for sources and more information. The paths also show the curves of maximum eclipse at two-minute increments in UT. These maps are also available on the web at *http://sunearth.gsfc.nasa.gov/eclipse/TSE2001/TSE2001.html*)

ESTIMATING TIMES OF SECOND AND THIRD CONTACTS

The times of second and third contact for any location not listed in this publication can be estimated using the detailed maps found in the final section. Alternatively, the contact times can be estimated from maps on which the umbral path has been plotted. Table 7 lists the path coordinates conveniently arranged in 1° increments of longitude to assist plotting by hand. The path coordinates in Table 3 define a line of maximum eclipse at five minute increments in time. These lines of maximum eclipse each represent the projection diameter of the umbral shadow at the given time. Thus, any point on one of these lines will witness maximum eclipse (i.e., mid-totality) at the same instant. The coordinates in Table 3 should be added to the map in order to construct lines of maximum eclipse.

The estimation of contact times for any one point begins with an interpolation for the time of maximum eclipse at that location. The time of maximum eclipse is proportional to a point's distance between two adjacent lines of maximum eclipse, measured along a line parallel to the center line. This relationship is valid along most of the path with the exception of the extreme ends, where the shadow experiences its largest acceleration. The center line duration of totality $\mathbf{D}$ and the path width $\mathbf{W}$ are similarly interpolated from the values of the adjacent lines of maximum eclipse as listed in Table 3. Since the location of interest probably does not lie on the center line, it is useful to have an expression for calculating the duration of totality $\mathbf{d}$ as a function of its perpendicular distance $\mathbf{a}$ from the center line:

$$\mathbf{d} = \mathbf{D} \ (1 - (2 \ \mathbf{a}/\mathbf{W})^2)^{1/2} \ \text{seconds} \qquad [3]$$

where: $\mathbf{d}$ = duration of totality at desired location (seconds)
$\mathbf{D}$ = duration of totality on the center line (seconds)
$\mathbf{a}$ = perpendicular distance from the center line (kilometers)
$\mathbf{W}$ = width of the path (kilometers)

If $\mathbf{t_m}$ is the interpolated time of maximum eclipse for the location, then the approximate times of second and third contacts ($\mathbf{t_2}$ and $\mathbf{t_3}$, respectively) are:

| Second Contact: | $t_2 = t_m - d/2$ | [4] |
| Third Contact: | $t_3 = t_m + d/2$ | [5] |

The position angles of second and third contact (either **P** or **V**) for any location off the center line are also useful in some applications. First, linearly interpolate the center line position angles of second and third contacts from the values of the adjacent lines of maximum eclipse as listed in Table 5. If X_2 and X_3 are the interpolated center line position angles of second and third contacts, then the position angles x_2 and x_3 of those contacts for an observer located **a** kilometers from the center line are:

| Second Contact: | $x_2 = X_2 - \arcsin(2\,a/W)$ | [6] |
| Third Contact: | $x_3 = X_3 + \arcsin(2\,a/W)$ | [7] |

where: x_n = interpolated position angle (either **P** or **V**) of contact **n** at location
X_n = interpolated position angle (either **P** or **V**) of contact **n** on center line
a = perpendicular distance from the center line (kilometers)
 (use negative values for locations south of the center line)
W = width of the path (kilometers)

MEAN LUNAR RADIUS

A fundamental parameter used in eclipse predictions is the Moon's radius k, expressed in units of Earth's equatorial radius. The Moon's actual radius varies as a function of position angle and libration due to the irregularity in the limb profile. From 1968 through 1980, the Nautical Almanac Office used two separate values for k in their predictions. The larger value ($k=0.2724880$), representing a mean over topographic features, was used for all penumbral (exterior) contacts and for annular eclipses. A smaller value ($k=0.272281$), representing a mean minimum radius, was reserved exclusively for umbral (interior) contact calculations of total eclipses [*Explanatory Supplement*, 1974]. Unfortunately, the use of two different values of k for umbral eclipses introduces a discontinuity in the case of hybrid or annular-total eclipses.

In August 1982, the International Astronomical Union (IAU) General Assembly adopted a value of $k=0.2725076$ for the mean lunar radius. This value is now used by the Nautical Almanac Office for all solar eclipse predictions [Fiala and Lukac, 1983] and is currently the best mean radius, averaging mountain peaks and low valleys along the Moon's rugged limb. The adoption of one single value for k eliminates the discontinuity in the case of annular-total eclipses and ends confusion arising from the use of two different values. However, the use of even the best 'mean' value for the Moon's radius introduces a problem in predicting the true character and duration of umbral eclipses, particularly total eclipses. A total eclipse can be defined as an eclipse in which the Sun's disk is completely occulted by the Moon. This cannot occur so long as any photospheric rays are visible through deep valleys along the Moon's limb [Meeus, Grosjean and Vanderleen, 1966]. But the use of the IAU's mean k guarantees that some annular or annular-total eclipses will be misidentified as total. A case in point is the eclipse of 3 October 1986. Using the IAU value for k, the *Astronomical Almanac* identified this event as a total eclipse of 3 seconds duration when it was, in fact, a beaded annular eclipse. Since a smaller value of k is more representative of the deeper lunar valleys and hence the minimum solid disk radius, it helps ensure the correct identification of an eclipse's true nature.

Of primary interest to most observers are the times when umbral eclipse begins and ends (second and third contacts, respectively) and the duration of the umbral phase. When the IAU's value for k is used to calculate these times, they must be corrected to accommodate low valleys (total) or high mountains (annular) along the Moon's limb. The calculation of these corrections is not trivial but must be performed, especially if one plans to observe near the path limits [Herald, 1983]. For observers near the center line of a total eclipse, the limb corrections can be more closely approximated by using a smaller value of k which accounts for the valleys along the profile.

This publication uses the IAU's accepted value of $k=0.2725076$ for all penumbral (exterior) contacts. In order to avoid eclipse type misidentification and to predict central durations which are closer to the actual durations at total eclipses, we depart from standard convention by adopting the smaller value of $k=0.272281$ for all umbral (interior) contacts. This is consistent with predictions in *Fifty Year Canon of Solar Eclipses: 1986 - 2035* [Espenak, 1987]. Consequently, the smaller k produces shorter umbral durations and narrower paths for total eclipses when compared with calculations using the IAU value for k. Similarly, predictions using a smaller k result in longer umbral durations and wider paths for annular eclipses than do predictions using the IAU's k.

LUNAR LIMB PROFILE

Eclipse contact times, magnitude and duration of totality all depend on the angular diameters and relative velocities of the Moon and Sun. Unfortunately, these calculations are limited in accuracy by the departure of the Moon's limb from a perfectly circular figure. The Moon's surface exhibits a rather dramatic topography, which manifests itself as an irregular limb when seen in profile. Most eclipse calculations assume some mean radius that averages high mountain peaks and low valleys along the Moon's rugged limb. Such an approximation is acceptable for many applications, but if higher accuracy is needed, the Moon's actual limb profile must be considered. Fortunately, an extensive body of knowledge exists on this subject in the form of Watts' limb charts [Watts, 1963]. These data are the product of a photographic survey of the marginal zone of the Moon and give limb profile heights with respect to an adopted smooth reference surface (or datum). Analyses of lunar occultations of stars by Van Flandern [1970] and Morrison [1979] have shown that the average cross-section of Watts' datum is slightly elliptical rather than circular. Furthermore, the implicit center of the datum (i.e., the center of figure) is displaced from the Moon's center of mass. In a follow-up analysis of 66,000 occultations, Morrison and Appleby [1981] have found that the radius of the datum appears to vary with libration. These variations produce systematic errors in Watts' original limb profile heights that attain 0.4 arc-seconds at some position angles. Thus, corrections to Watts' limb data are necessary to ensure that the reference datum is a sphere with its center at the center of mass.

The Watts charts have been digitized by Her Majesty's Nautical Almanac Office in Herstmonceux, England, and transformed to grid-profile format at the U. S. Naval Observatory. In this computer readable form, the Watts limb charts lend themselves to the generation of limb profiles for any lunar libration. Ellipticity and libration corrections may be applied to refer the profile to the Moon's center of mass. Such a profile can then be used to correct eclipse predictions which have been generated using a mean lunar limb.

Along the path, the Moon's topocentric libration (physical + optical) in longitude ranges from l=-3.1° to l=-4.6°. Thus, a limb profile with the appropriate libration is required in any detailed analysis of contact times, central durations, etc.. But a profile with an intermediate value is useful for planning purposes and may even be adequate for most applications. The lunar limb profile presented in Figure 8 includes corrections for center of mass and ellipticity [Morrison and Appleby, 1981]. It is generated for 13:00 UT, which corresponds to western Zambia near the border with Angola. The Moon's topocentric libration is l=-4.39°, and the topocentric semi-diameters of the Sun and Moon are 944.3 and 988.0 arc-seconds, respectively. The Moon's angular velocity with respect to the Sun is 0.364 arc-seconds per second.

The radial scale of the limb profile in Figure 8 (at bottom) is greatly exaggerated so that the true limb's departure from the mean lunar limb is readily apparent. The mean limb with respect to the center of figure of Watts' original data is shown (dashed) along with the mean limb with respect to the center of mass (solid). Note that all the predictions presented in this publication are calculated with respect to the latter limb unless otherwise noted. Position angles of various lunar features can be read using the protractor marks along the Moon's mean limb (center of mass). The position angles of second and third contact are clearly marked along with the north pole of the Moon's axis of rotation and the observer's zenith at mid-totality. The dashed line with arrows at either end identifies the contact points on the limb corresponding to the northern and southern limits of the path. To the upper left of the profile are the Sun's topocentric coordinates at maximum eclipse. They include the right ascension *R.A.*, declination *Dec.*, semi-diameter *S.D.* and horizontal parallax *H.P.*. The corresponding topocentric coordinates for the Moon are to the upper right. Below and left of the profile are the geographic coordinates of the center line at 13:00 UT while the times of the four eclipse contacts at that location appear to the lower right. Directly below the profile are the local circumstances at maximum eclipse. They include the Sun's altitude and azimuth, the path width, and central duration. The position angle of the path's northern/southern limit axis is *PA(N.Limit)* and the angular velocity of the Moon with respect to the Sun is *A.Vel.(M:S)*. At the bottom left are a number of parameters used in the predictions, and the topocentric lunar librations appear at the lower right.

In investigations where accurate contact times are needed, the lunar limb profile can be used to correct the nominal or mean limb predictions. For any given position angle, there will be a high mountain (annular eclipses) or a low valley (total eclipses) in the vicinity that ultimately determines the true instant of contact. The difference, in time, between the Sun's position when tangent to the contact point on the mean limb and tangent to the highest mountain (annular) or lowest valley (total) at actual contact is the desired correction to the predicted contact time. On the exaggerated radial scale of Figure 8, the Sun's limb can be represented as an epicyclic curve that is tangent to the mean lunar limb at the point of contact and departs from the limb by **h** through:

$$\mathbf{h} = \mathbf{S} \ (\mathbf{m}-1) \ (1-\cos[\mathbf{C}]) \qquad\qquad [8]$$

where: $\mathbf{h}$ = departure of Sun's limb from mean lunar limb
$\mathbf{S}$ = Sun's semi-diameter
$\mathbf{m}$ = eclipse magnitude
$\mathbf{C}$ = angle from the point of contact

Herald [1983] has taken advantage of this geometry to develop a graphical procedure for estimating correction times over a range of position angles. Briefly, a displacement curve of the Sun's limb is constructed on a transparent overlay by way of equation [8]. For a given position angle, the solar limb overlay is moved radially from the mean lunar limb contact point until it is tangent to the lowest lunar profile feature in the vicinity. The solar limb's distance $\mathbf{d}$ (arc-seconds) from the mean lunar limb is then converted to a time correction Δ by:

$$\Delta = \mathbf{d} \ v \ \cos[\mathbf{X} - \mathbf{C}] \qquad\qquad [9]$$

where: Δ = correction to contact time (seconds)
$\mathbf{d}$ = distance of Solar limb from Moon's mean limb (arc-sec)
v = angular velocity of the Moon with respect to the Sun (arc-sec/sec)
$\mathbf{X}$ = center line position angle of the contact
$\mathbf{C}$ = angle from the point of contact

This operation may be used for predicting the formation and location of Baily's beads. When calculations are performed over a large range of position angles, a contact time correction curve can then be constructed.

Since the limb profile data are available in digital form, an analytical solution to the problem is possible that is quite straightforward and robust. Curves of corrections to the times of second and third contact for most position angles have been computer generated and are plotted in Figure 8. The circular protractor scale at the center represents the nominal contact time using a mean lunar limb. The departure of the contact correction curves from this scale graphically illustrates the time correction to the mean predictions for any position angle as a result of the Moon's true limb profile. Time corrections external to the circular scale are added to the mean contact time; time corrections internal to the protractor are subtracted from the mean contact time. The magnitude of the time correction at a given position angle is measured using any of the four radial scales plotted at each cardinal point.

For example, Table 15 gives the following data for Lusaka, Zambia:

Second Contact = 13:09:19.3 UT P_2=118°
Third Contact   = 13:12:32.8 UT P_3=247°

Using Figure 8, the measured time corrections and the resulting contact times are:

C_2=+4.0 seconds;   Second Contact = 13:09:19.3 +4.0s   = 13:09:23.3 UT
C_3=−1.2 seconds;   Third Contact   = 13:12:32.8 −1.2s   = 13:12:31.6 UT

The above corrected values are within 0.2 seconds of a rigorous calculation using the actual limb profile.

Lunar limb profile diagrams for a number of other positions/times along the path of totality are available via a special web site of supplemental material for the total solar eclipse of 2001 (*http://sunearth.gsfc.nasa.gov/eclipse/TSE2001/TSE2001.html*).

LIMB PROFILE EFFECTS ON THE DURATION OF TOTALITY

As was previously discussed, the Moon's center of figure (i.e., the geometric center of the Watts' datum) is displaced from the Moon's center of mass. A case in point is the lunar limb geometry at 13:00 UT (Figure 8) where the center of figure is displaced −0.13 arc-seconds in ecliptic latitude and +0.53 arc-seconds in ecliptic longitude. This shift is fairly characteristic along much of the 2001 umbral path but varies considerably between eclipses due to different libration geometry's. Since most predictions appearing in this publication are calculated with respect to the Moon's center of mass, the center of figure offset has a small but significant consequence on the duration of totality. When compounded with the irregularities of the lunar limb profile, the overall result is to shift the maximum duration of totality north of the center line by 0-9 kilometers along the African path, and 10 kilometers north along the Madagascar path.

Figure 9 shows a series of calculations for the duration of totality within ±60 kilometers of the center line and spaced at ten minute intervals along the path through Africa and Madagascar. For a given time, the duration of totality is calculated at 1 kilometer intervals perpendicular to the path within a 120 kilometer zone centered on the center line. Predictions using the Moon's center of mass and mean limb are represented by the dotted curves. Predictions using the actual limb profile to calculate corrected contact times and the resulting duration of totality are plotted as solid curves. What becomes immediately apparent upon inspection of Figure 9, is the asymmetry of the true limb duration curves and is a consequence of the complex Sun/Moon limb geometry which changes quickly with path position.

Observers wishing to witness the maximum possible duration of totality from a given section of the path can use Figure 9 to optimize their location with respect to the center line.

LIMB CORRECTIONS TO THE PATH LIMITS: GRAZE ZONES

The northern and southern umbral limits provided in this publication were derived using the Moon's center of mass and a mean lunar radius. They have not been corrected for the Moon's center of figure or the effects of the lunar limb profile. In applications where precise limits are required, Watts' limb data must be used to correct the nominal or mean path. Unfortunately, a single correction at each limit is not possible since the Moon's libration in longitude and the contact points of the limits along the Moon's limb each vary as a function of time and position along the umbral path. This makes it necessary to calculate a unique correction to the limits at each point along the path. Furthermore, the northern and southern limits of the umbral path are actually paralleled by a relatively narrow zone where the eclipse is neither penumbral nor umbral. An observer positioned here will witness a slender solar crescent that is fragmented into a series of bright beads and short segments whose morphology changes quickly with the rapidly varying geometry between the limbs of the Moon and the Sun. These beading phenomena are caused by the appearance of photospheric rays that alternately pass through deep lunar valleys and hide behind high mountain peaks as the Moon's irregular limb grazes the edge of the Sun's disk. The geometry is directly analogous to the case of grazing occultations of stars by the Moon. The graze zone is typically five to ten kilometers wide and its interior and exterior boundaries can be predicted using the lunar limb profile. The interior boundaries define the actual limits of the umbral eclipse (both total and annular) while the exterior boundaries set the outer limits of the grazing eclipse zone.

Table 6 provides topocentric data and corrections to the path limits due to the true lunar limb profile. At five minute intervals, the table lists the Moon's topocentric horizontal parallax, semi-diameter, relative angular velocity of the Moon with respect to the Sun and lunar libration in longitude. The Sun's center line altitude and azimuth is given, followed by the azimuth of the umbral path. The position angle of the point on the Moon's limb which defines the northern limit of the path is measured counter-clockwise (i.e., eastward) from the north point on the limb. The path corrections to the northern and southern limits are listed as interior and exterior components in order to define the graze zone. Positive corrections are in the northern sense while negative shifts are in the southern sense. These corrections (minutes of arc in latitude) may be added directly to the path coordinates listed in Table 3. Corrections to the center line umbral durations due to the lunar limb profile are also included and they are mostly positive. Thus, when added to the central durations given in Tables 3, 4, 5 and 7, a slightly longer central total phase is predicted.

Detailed coordinates for the zones of grazing eclipse at each limit for all land based sections of the path are presented in Table 8. Given the uncertainties in the Watts data, these predictions should be accurate to ±0.3 arc-seconds. The interior graze coordinates take into account the deepest valleys along the Moon's limb which produce the simultaneous second and third contacts at the path limits. Thus, the interior coordinates define the true edge of the path of totality. They are calculated from an algorithm which searches the path limits for the extreme positions where no photospheric beads are visible along a ±30° segment of the Moon's limb, symmetric about the extreme contact points at the instant of maximum eclipse. The exterior graze coordinates are somewhat arbitrarily defined and calculated for the geodetic positions where an unbroken photospheric crescent of 60° in angular extent is visible at maximum eclipse.

In Table 8, the graze zone latitudes are listed every 30' in longitude (at sea level) and include the time of maximum eclipse at the northern and southern limits as well as the path's azimuth. To correct the path for locations above sea level, *Elev Fact*[8] is a multiplicative factor by which the path must be shifted north perpendicular to itself (i.e., perpendicular to path azimuth) for each unit of elevation (height) above sea level. To calculate the shift, a location's elevation is multiplied by the *Elev Fact* value. Positive values (usually the case for eclipses in the Southern Hemisphere) indicate that the path must be shifted north. For instance, if one's elevation is 1000 meters above sea level and the *Elev Fact* value is +0.20, then the shift is +200m (= 1000m x +0.20). Thus, the observer must shift the path coordinates 200 meters in a direction perpendicular to the path and in a negative or southerly sense.

---

[8] The elevation factor is the product, tan(90-A) * sin(D), where A is the altitude of the Sun and D is the difference between the azimuth of the Sun and the azimuth of the limit line, with the sign selected to be positive if the path should be shifted north with positive elevations above sea level.

The final column of Table 8 lists the *Scale Fact* (km/arc-second). This scaling factor provides an indication of the width of the zone of grazing phenomena, due to the topocentric distance of the Moon and the projection geometry of the Moon's shadow on Earth's surface. Since the solar chromosphere has an apparent thickness of about 3 arc-seconds, and assuming a *Scale Fact* value of 2 km/arc-seconds, then the chromosphere should be visible continuously during totality for any observer in the path who is within 6 kilometers (=2x3) of each interior limit. However, the most dynamic beading phenomena occurs within 1.5 arc-seconds of the Moon's limb. Using the above Scale Factor, this translates into the first 3 kilometers inside the interior limits. But observers should position themselves at least 1 kilometer inside the interior limits (south of the northern interior limit or north of the southern interior limit) in order to ensure that they are inside the path due of to small uncertainties in Watts' data and the actual path limits.

For applications where the zones of grazing eclipse are needed at a higher frequency in longitude interval, tables of coordinates every 7.5' in longitude are available via a special web site for the total solar eclipse of 2001 (*http://sunearth.gsfc.nasa.gov/eclipse/TSE2001/TSE2001.html*).

SAROS HISTORY

The periodicity and recurrence of solar (and lunar) eclipses is governed by the Saros cycle, a period of approximately 6,585.3 days (18 years 11 days 8 hours). When two eclipses are separated by a period of one Saros, they share a very similar geometry. The eclipses occur at the same node with the Moon at nearly the same distance from Earth and at the same time of year. Thus, the Saros is useful for organizing eclipses into families or series. Each series typically lasts 12 to 13 centuries and contains 70 or more eclipses.

The total eclipse of 2001 June 21 is the fifty-seventh member of Saros series 127 (Table 16), as defined by van den Bergh [1955]. All eclipses in the series occur at the Moon's ascending node and the Moon moves southward with each member in the family (i.e. - gamma[9] decreases). The series is a mature one which began with a small partial eclipse at high northern hemisphere latitudes on 0991 Oct 10. After twenty partial eclipses each of increasing magnitude, the first central eclipse occurred on 1352 May 14. The event was a two minute total eclipse with a path sweeping through Greenland.

The series continued to produce total eclipses of increasing duration as the path of each event shifted further south. The fifth total eclipse is noteworthy because it passed through northern Europe on 1406 Jun 16. "It was so dark that...people could not recognize one another...(they) thought that the end of the world was coming." [Stephenson, 1997]. As subsequent events occurred, the duration continued to grow to a maximum of 5 minutes 40 seconds on 1532 Aug 30. As a consequence to the path geometry across Earth's surface, the duration of ensuing eclipses began to fall but remained above four minutes.

The paths reversed their southern migration and drifted northward during the 18th and 19th centuries. This effect occurred as a result of Earth's passage from winter solstice through spring equinox whereby the northern hemisphere rapidly tipped towards the Sun. During this period, the duration again approached the five minute mark and exceeded it on 1929 May 9. The umbral path crossed Malaysia and the Philippines. The track of the following eclipse stretched from South America through equatorial Africa on 1947 May 20. The twenty-fifth total eclipse of saros 127 traversed the South Pacific on 1965 May 30.

The most recent eclipse occurred on 1983 June 11. Indonesians witnessed a maximum total phase of 5 minutes 11 seconds. After 2001, the following member occurs on 2019 Jul 02. It's track crosses the Pacific and South America (*http://sunearth.gsfc.nasa.gov/eclipse/eclipse/SEplot/SE2019Jul02.gif*) .

The central eclipse phase of saros 127 comes to an end with the total eclipse of 2091 Aug 15. Its landless track falls over open ocean between Australia and Antarctica. The umbral shadow misses Earth's surface during a deep partial eclipse on 2109 Aug 26. The eclipse magnitude decreases steadily during the next three centuries. Finally, the saros ends with a brief partial eclipse on 2452 Mar 21. A detailed list of eclipses in Saros series 127 appears in Table 8. For a more detailed list including local circumstances at greatest eclipse, see: *http://sunearth.gsfc.nasa.gov/eclipse/SEsaros/SEsaros127.html*

In summary, Saros series 127 includes 82 eclipses. It begins with 20 partials, followed by 42 total eclipses and ends with 20 more partials. The total duration of Saros 127 is 1460.44 years.

---

[9] Minimum distance of the Moon's shadow axis from Earth's center in units of equatorial Earth radii. Gamma defines the instant of greatest eclipse and takes on negative values south of the Earth's center.

WEATHER PROSPECTS FOR THE ECLIPSE

INTRODUCTION

The eclipse occurs while southern Africa is in the midst of its winter dry season. Fine sunny weather dominates the sky in mid June (the second sunniest region of the globe at this time) and the eclipse seeker who ventures to the most favorable parts of the shadow track has excellent prospects for success.

OVERVIEW

THE GEOGRAPHY

The interior of southern Africa is a vast plateau lying about 1000 m above sea level with a cool and pleasant winter climate. Surrounding this plateau is a jumble of hilly and mountainous terrain ending abruptly in a narrow coastal belt that hugs the Atlantic Ocean on one side and the Indian Ocean on the other. On the Atlantic side of the African continent the cool Benguela current bathes the coast of Angola with 13 degree temperatures, moderating the heat and cloudiness. On the eastern side the wide Equatorial Current of the South Indian Ocean meets the African coast near the Tanzania-Mozambique border. This warm 20° current then turns southward to flow through the Mozambique Channel west of Madagascar.

Over Zimbabwe, the mountain barrier that separates the broad coastal lowlands of Mozambique from the plateau is known as the Eastern Highlands. Reaching an altitude of more than 2000 m, the Highlands block the flow of moist easterly trade winds from the Indian Ocean and the Mozambique Channel. Further north, between Malawi and Mozambique, the Mlanje Mountains serve the same function.

The eastern mountain barrier is broken in two places. To the south, between Zimbabwe and South Africa, the Limpopo River valley provides weather systems with convenient passage into Botswana and eventually to southern Zambia near Victoria Falls (Figure 10). Further north, the Zambezi River provides an opening to the interior along the Mozambique-Zimbabwe border, occasionally bringing cloudy skies as far inland as Lusaka in Zambia. This second entrance to the continent is of most importance to eclipse watchers because the track of the shadow runs more-or-less along the course of the Zambezi through Zimbabwe and Mozambique.

Over Madagascar the terrain is more dramatic. A prominent mountain ridge extends the length of the island, ascending over 2500 m in some places. On the sharply rising eastern side the prevailing easterly trade winds are forced to rise upward, depositing much of their moisture on the slopes. Once across the divide however, the descending airflow is rapidly dried, bringing semi-arid conditions to the southwest corner of the island where the eclipse track comes ashore.

CLIMATE CONTROLS

The eclipse track crosses the African continent just to the north of a semi-permanent high pressure belt that circles the globe at about 30° south latitude (Figure 10). In June this anticyclonic zone straddles the southern tip of South Africa, with one center in the Indian Ocean off Madagascar and another in the Atlantic to the west of Namibia. These high pressure centers represent the polar side of the southern Hadley circulation, a zone of notably sunny skies and limited precipitation. Only the Sahara Desert can claim less cloud cover in the month of June than the western parts of the African eclipse track.

The Intertropical Convergence Zone (ICZ) is Earth's "weather equator" and is formed by the gentle collision of air flows from the northern and southern hemispheres. It is a region of heavy cloudiness, high relative humidity and considerable precipitation, but at this time of year, in the southern winter, it tends to be found well away from the eclipse track, on the other side of the equator at about 20° north (Figure 11). Occasionally the tropical air associated with the ICZ will make an excursion into western Angola, but for the most part it stays well away from the eclipse track during the southern hemisphere's winter.

The oceans on each side of the continent are other more obvious moisture sources. On the Atlantic side, only small amounts of cloud and fog occasionally drift onto the Angolan shores. The story is not the same in the east however, where low level cloud and light precipitation are frequent visitors to the

Mozambique lowlands and the slopes of the Eastern Highlands of Zimbabwe. Still other cloudiness comes with the passage of weak frontal systems, though these are modest in size in comparison with the continent-straddling weather systems of Europe and North America. The fronts arrive from the south, temporarily breaking through the anticyclonic belt, and bringing a notable change in the weather when they reach the mountainous parts of Zimbabwe and Mozambique.

Much of the region of the eclipse track has an annual precipitation in excess of 1000 mm, divided between a summer wet season and a winter dry. The dry season begins abruptly in May and by the time of the eclipse, is in full sway. June precipitation amounts to only a few millimeters in many locations, and in much of Zambia and Angola, is less than one millimeter. The dry season, when it arrives, does so with great authority.

THE WEATHER IN DETAIL

ANGOLA

In Angola, a narrow coastal plain rises rapidly to a broad plateau lying above 1000 m altitude. The two regions (plain and plateau) have distinct temperature regimes, but are equally blessed with fine eclipse skies in June. During winter there is little flow from the Atlantic over the Angolan plateau and skies are clear with little or no precipitation. While climatological statistics are sparse due to the decades-long civil war, measurements from satellite suggest that the best skies along the eclipse track are to be expected on the plateau. The mean cloudiness in this region is less than 10% and occasionally below 5% (Figure 11).

The only station record of sunshine close to the eclipse path comes from Huambo, just outside totality on the western side of the plateau. The nine hours of sunshine there (79% of the total possible) is not especially noteworthy in comparison with other sites in Zambia and Zimbabwe (Table 17), but other indicators, particularly the satellite measurements of cloudiness, show that interior Angola is a fine location from a meteorological point of view. Zambian sites close to the Angolan border (especially Mongu) have recorded as much as 9.9 hours of sunshine per day (~90% of the maximum possible) and so it would seem reasonable that central Angolan sites may exceed 10 hours of daily sunshine, likely in the area near Andulo.

Satellite images for June and early July of 1998 show only an occasional patch of fog or low cloud drifting onto the coastal lowlands from the Benguela Current (Figure 13). The same images showed virtually no cloud over the interior during a 35-day period. However, an orbital study of biomass burning (Figure 12) shows that the track through Angola passes through a region with a large number of seasonal fires. A smoky haze is likely to hang over the region, possibly making the outer parts of the corona more difficult to see and photograph.

Much of this is academic, as the ongoing Angolan civil war makes travel into the interior hazardous, in spite of a UN-brokered cease fire.

ZAMBIA

The eclipse track crosses into Zambia at Chavuma Falls, passing north of the city of Zambezi. From here it heads east southeastward across Kafue National Park to pass 40 km north of the capital of Lusaka, the only major city to lie within totality. East of Lusaka, the umbral path reaches the Zambezi River again, passing just to the north of Lake Kariba to cross into Zimbabwe.

Access to the center line is relatively easy in Zambia. The highway from Lusaka west to Mongu (M9) stays within the umbral shadow for over 300 km, though well to the south of the centerline. This route is described as being in "protracted repair" by one guide book, but is the main route to Kafue National Park and the western half of the country. Two other highways head north from Lusaka to cross the centerline in two places. Landless Corner, 69 km north of the capital, lies just beyond the middle of the shadow path on the Great North Road. To the northeast, the Great East Road to Chipata, one of the best in the country, crosses the centerline at about 60 km distance.

Zambia has a number of safari settings for the eclipse, particularly at Kafue National Park in the northwestern part of the country. This park is one of the largest in Africa, swampy in the wet season but offering some of the best weather along the track in June. The Kafue Park gate is easy to reach from Lusake, 279 km to the east, but the centerline lies a further hundred km to the north of the gate and must be traveled on softer roads within the park. Even though it is the dry season, 4-wheel drive vehicles may be necessary to reach the shadow axis by this route.

There are over 70 different tribes in Zambia, but one has a special association with eclipses. Originally from South Africa, the Ngoni migrated into eastern Zambia (and into Zimbabwe) around 1830 to escape the Great Shaka Zulu. A solar eclipse and a great celebration marked the day they crossed the Zambezi to their new homeland. This eclipse and the one in 2002 which also touches Zambia and Zimbabwe will likely provide opportunity for further celebration.

Winter months in Zambia are virtually rainless. The circulation associated with the anticyclones to the south dries the mid levels of the atmosphere and suppresses any tendency to form rain-bearing clouds. The southeast trade winds, blowing first across Madagascar and then Mozambique, carry little moisture into the mid-continent in June, blocked in large part by the Eastern Highlands.

The most important source of cloudiness in Zambia in June comes in association with the passage of a succession of highs and lows across South Africa at intervals of a few days. Passage of a high pressure cell results in a flow of cold air from the south that is restricted in its inland motion by the Drakensberg Mountains in South Africa and the Eastern Highlands in Zimbabwe. The cold air finds an opening into the continent along the Limpopo River, and less commonly, along the Zambezi. This air has a high relative humidity and as it rises along the slopes of the Eastern Highlands and the Zambezi Escarpment, pushed by easterly winds, it quickly turns into an extensive broken stratiform cloudiness. Much of the impact of this cloudiness is felt over the lowlands of Mozambique and eastern Zimbabwe, but the continuing push of the winds frequently forces the cloud deep into the Limpopo and Zambezi valleys.

This pattern of cloudiness is known locally as a *guti*; its counterpart in Malawi is the *chiperoni*. The cool southerly winds, cloudiness, and drizzle can bring an abrupt change to the normally sunny and dry weather. These invasions of cold air occur year round, at about weekly intervals, but in Zambia, well inland from the Mozambique Channel, the incidence of cloudiness is much reduced (Figure 13) and the cloud cover tends to be more scattered than at lower elevations in the Zambezi Valley. It is uncommon for the *guti* to have an influence beyond a line from Lusaka to Mpika, along the Great North Road. In the graphs of Figure 13 the effect can be seen by comparing the cloudiness in "Lusaka to Kafue Park" with "Lusaka to the Zimbabwe Border". The effect of the *guti* can also be seen in the sunshine statistics in Table 17.

Moisture that travels inland from the south, following the Limpopo River, is unlikely to make it as far north as the eclipse track. The sunshine statistics for Livingstone, Victoria Falls and Bulawayo speak to the efficiency of the blocking caused by the Eastern Highlands.

A *guti* usually doesn't persist for more than a day in the vicinity of Lusaka, breaking up quickly as it moves onto the plateau. A high sun and generally drier surrounding conditions soon change the stratiform cloud into cumuliform types that mix with the air of the plateau and dissipate. Satellite images from 1998 show that patchy cloudiness associated with a *guti*, cloud that might affect observations of the eclipse, was observed in the Lusaka-Zambezi area on seven occasions in the 35 days of observation (Figure 13). Twenty-eight days were clear or had only scattered cloud and would not have interfered with a view of the Sun.

Of the seven critical days, three had scattered to broken cloud that would likely disappear between first and second contact, though some movement might be necessary to avoid the larger patches. The remaining four days would have required more definite movement, perhaps a hundred kilometers to the west, though on two of these occasions a patch of cloud drifted west almost to Kafue National Park.

On days that will tend to be cloudy, the morning often greets the sunrise with a heavy cloud cover, sometimes overcast. As the sun rises, the cloud will dissipate, being relatively thin and at a high level so that it mixes readily with drier atmospheric layers. The improvement doesn't last however, as a little more heating brings the onset of cumulus cloudiness, encouraged by the invisible but still-present moisture which formerly fed the dawn cloudiness. Thus we have a very characteristic pattern of morning cloud, promising mid-morning clearing, and then a return to noontime cloudiness again.

If the morning cloud burns off quickly and the resulting noontime cumulus clouds are widely dispersed, then the cooling associated with the arrival of the shadow is most likely to dissipate the small amount of cloud and bring a successful eclipse. If the noon clouds are heavy and return the sky to an near-overcast condition, then the eclipse is in danger. Such skies are likely to be reluctant to clear as the shadow approaches, and instead settle into a stratiform layer that will obscure much of the spectacle. Even in this case the whole event may not be lost, as central African cloud layers are typically very thin in this season and the inner parts of the corona and the prominences may be quite visible. Nevertheless, a heavy cumulus cloudiness at noon is a sign that alternative plans should be readied for use.

Viewing locations along the Great East Road, which runs parallel to and about 50 km west of the Zambezi, are most at risk. Selecting a backup site along the Great North Road (though Zambezi cloudiness can reach this area as well) would be a good idea for those planning to choose locations east of Lusaka. Fortunately, the approach of cloud along the Zambezi River can often be seen a day or two ahead from satellite imagery.

As in Angola, one element that will likely affect the view of the eclipse is the extensive haziness that overlies the interior plateau in the winter months (Table 17, Figure 12). This haziness is man-made, the consequence of an extensive burning of the veldt that begins in earnest in June and reaches its worst in September or October. Much of the burning occurs in the grasslands north of the eclipse track and occasionally visibilities may be reduced below 5 km. The haze is thickest in the morning hours and disperses through the lower atmosphere during the day. While unlikely to have a major impact on the view of the eclipse, it could reduce the visibility of the fainter parts of the outer corona.

Temperatures on the Zambia-Zimbabwe plateau are pleasant but cool in June, with daytime highs generally in the low to mid twenties and morning lows between 5 and 10 degrees. The dry air and generally cloudless skies should bring a significant drop in ground temperatures during the eclipse especially on the plateau west of Lusaka. Occasional winter cold outbreaks come overland from Botswana and arrive in Zambia and Zimbabwe with dew points well below zero. This dry air brings no cloudiness with it, but nighttime temperatures can fall below the freezing mark in southern parts and below the 5° mark near the eclipse track.

ZIMBABWE

After crossing the Zambezi, the umbral path is shared by both Zimbabwe and Mozambique for 400 kilometers until it passes entirely into Mozambique near Changara. The south side of the track and most of the centerline belongs to Zimbabwe, while the north half crosses the length of western Mozambique. In Zimbabwe the track traverses the lip of the Zambezi escarpment, looking down on Lago de Cahora Bassa in Mozambique. The heights of the escarpment are quite substantial, rising more than 1500 m above the river valley in the Mavuradonha Mountains north of Centenary.

Zimbabwe has a well-developed tourist infrastructure and accommodation for a large number of people. Mount Darwin and Centenary, both within the path and well connected to Harare by road, are likely to prove popular setting-off points for the center line. According to the Michelin map, good roads lead north from both of these communities, reaching to within a few kilometers of the shadow axis. Geoffrey Carew of Carew Safaris (Harare) reports that "roads to the center line are good quality tar as far as Muzarabani and then reasonable dirt roads from there on."

There are numerous safari camps across northern Zimbabwe, some very close to or even on the centerline, which will allow a relaxing and very exotic wait for the Moon's shadow. Most of these camps can hold only a small number of people, but at least some have plans to make accommodation for larger groups for the eclipse. In particular, the Mavuradonha Wilderness, a game reserve of rugged uplands on the Escarpment, may prove to be an irresistible destination for groups of eclipse watchers who might enjoy the African outdoor experience. Carew Safaris is planning a riding safari to view the eclipse at 1660 m from Makura Mountain, an altitude that should improve the weather prospects even further.

The path enters Zimbabwe just north of Mana Pools National Park. The Park is designed mostly for canoe or walking safaris. According to Trish Berry of the Zambezi Safari and Travel Company (Kariba), Mana Park "is not usually an area that one can drive through independently" as specific licenses and professional guides are required for a visit. There is a minimum of nine persons per guide. Visitors in this area will have to go extra trouble to reach the centerline and scrounge the last few seconds out of totality. This may be a location where it is better to sacrifice the additional time in exchange for a magnificent safari setting, or become more adventuresome and take a canoe trip down the Zambezi to maximum eclipse.

Perhaps the greatest limitation to Zimbabwean sites is the lack of convenient routes for movement to sunnier locations should the selected site turn suddenly cloudy, as all roads tend to run across rather than along the track. The easiest highway track to the eclipse lies on the route from Harare to Tete (a distance of 238 kilometers to the centerline). The actual centerline lies slightly inside Mozambique, just beyond the border crossing at Nyamapanda, but the loss in the duration of totality from the Zimbabwean side is of the order of only a few seconds.

For the most part, the weather through northern Zimbabwe is similar to that in eastern Zambia. Cloudiness is only slightly higher, and the percent of possible sunshine correspondingly lower, by a few percentage points. June precipitation is higher than in Zambia, but barely so (less than 1 mm at Mount Darwin), at least in the vicinity of the shadow track. Because the Zimbabwean half of the track lies on the heights of the Zambezi escarpment, it escapes the larger part of the cloud that traverses the valley below and which gives a distinctly higher cloud cover to the Mozambican half of the track.

From a climatological point of view, the best sites in Zimbabwe will be those in the west, close to the Zambia border. The general trend to increasing cloudiness as the track moves eastward is reflected in the satellite measurements in Figure 11 and the statistics in Table 17. In the west near the Zambia border,

Kariba and Karoi, both just outside the path of totality, show 82 and 76% of possible sunshine respectively. Slightly farther along the track, Mount Darwin reports 74% of the maximum possible, while Tete (Mozambique), much lower in the valley and farther east, reports a meager 58%. These gradual trends in cloudiness are confirmed by the behavior of the cloud patterns in 1998 graphed in Figure 13. The high levels of cloudiness at Tete do not recommend a location near Nyamapanda unless a favorable forecast for the area can be obtained a day or two ahead of time for the area.

As in Zambia, temperatures in Zimbabwe are pleasant and humidity is low, except during a *guti*. The frequency of hazy skies is considerably lower than in Zambia, evidently because the extent of biomass burning is lower than farther west (Figure 12).

MOZAMBIQUE

Mozambique is the cloudiest location from which to attempt to see the eclipse, but access to the track is relatively easy in western sections. One location, on the highway from Blantyre in Malawi to Harare, has already been mentioned in the section on Zimbabwe. The highway passes through Mozambique for most of the journey, along a route referred to as the Tete corridor. Proceeding from Blantyre, the highway first crosses the Zambezi at Tete over an impressive suspension bridge and then proceeds southward to Changara where it forks into southerly and southwesterly branches. Both forks cross the center line about 40 kilometers farther along. The highway through the Tete corridor is well traveled and in good condition, allowing easy access to the eclipse.

Farther east near the delta of the Zambezi, the highway from Beira to Quelimane reaches the centerline just south of the river. This appears to be a more difficult route as sections of the highway are under repair, and so the access to the track in the Tete corridor is likely to appeal to most travelers, especially those coming from the direction of Malawi.

June is the cool season in Mozambique, especially inland, though the climate is moderated by the easterly flow from the twenty-degree waters of the Mozambique Channel. The rainy season ends by April in the interior but may linger as long as June on some parts of the coast, especially over the Zambezi delta where the eclipse track exits the continent. And though it is 430 km away to the east, Madagascar has some considerable influence on the weather on the mainland. The southeasterly trade winds flowing across the island lose much of their moisture on the windward side and can recover only a part of the loss in the short trajectory over the Channel before reaching the coast of Mozambique.

The climate of Mozambique is primarily a lowland one, with higher precipitation, temperature, and cloud cover than the inland plateau. Cloud statistics in Table 17 favor areas near the coast, with lighter cloud in Quelimane and Beira than in Tete. Average sunshine is 8.4 hours per day at Beira, on the coast south of the track, but declines to 6.5 hours per day at Tete, 420 km inland. Satellite images show that most cloud in the area seems to form over the land, with much sunnier conditions over the waters along the coast. This is likely the reason that Quelimane and Beira have a greater amount of sunshine than Tete. However getting to a site on the centerline near the coast requires travel to Chinde from Quelimane for about 80 km along a poor road and there appears to be several small rivers or streams to ford along the way according to the Michelin map.

An alternative route has access from the Beira-Quelimane highway, turning south to Mopela, and then traveling along the north shore of the Zambezi for about 30 km to reach the center line. This location is best approached from the Quelimane side, as the road from Beira is under extensive reconstruction.

Temperatures along the lowlands of Mozambique are similar to the highlands but relative humidities average 10% higher. Winds blow steadily from the southeast in the winter, with small variations due to the effect of the terrain. Biomass burning is subdued in comparison with that on the plateau (Figure 12), and is concentrated around Beira and Maputo, sparing the shadow track south of Quelimane.

Quelimane has a very high frequency of fog overnight (over 25% in June) and may show a tendency to cloud over at the last moment before totality. This poses a bit of a dilemma for those willing to experiment with a beach site under the track near Chinde, since the favorable cloud climatology is partly undone by the potential for fog.

MADAGASCAR

Exotic Madagascar, lying in the path of the Indian Ocean trade winds, would seem to be poorly served for the eclipse observer, but its mountainous terrain blocks the moisture-laden easterly flow and forms a delightful pocket of clear skies on the southwest coast. Moisture-bearing winds, when forced to rise by topography, cool and become saturated. Clouds form (heavy cloudiness if there is enough moisture) and precipitation becomes steady and widespread. The windward side of an ocean-facing mountain chain usually develops immense rain forests from the steady supply of liquid, though the occurrence of cloud and rain may be moderated by dry and wet seasons.

The counterpart of the moist upslope flow is the descending airflow that comes on the lee side of the peaks. Just as the upward flow becomes cloudy, so the downward becomes dry and sunny. The drying can be very dramatic, as it is in Madagascar, with tropical rain forests being replaced by semi-arid vegetation on the western side of the island. Climatological records and satellite images mark this western coast as the second best eclipse site along the track, bested only by western Zambia and Angola.

Southwest Madagascar near Toliara is known for its spiny forests, cactus-like trees inhabited by exotic birds, and for the Mahafaly, Masikoro and Bara tribes, famous for the design of their tombs. The Sun is only 12 degrees above the horizon at totality and actually sets moments before the eclipse ends at fourth contact. Observers here will be treated to an image of a fiery red orb with a tiny bite taken out, setting over the ocean horizon to the west. Perhaps a green flash will complement the view!

The center of the eclipse comes ashore north of Morombe, into a landscape of dry scrub forest and occasional rivers. The town, small and dusty with a bustling beach waterfront, lies 306 km north of Toliara, one of Madagascar's major cities. The road from Toliara to Morombe is slow and grating, being composed of the broken remains of a formerly paved highway that has aged into a mixture of sandy potholes and abrupt edges. Speeds will be slow—perhaps only 30 km/h or so—and so the trip between the two communities takes the better part of one or two days depending on the mode of transportation. An alternative route runs northward from Toliara along the coast, but this is a convoluted single-lane sand trail which is only passable by four-wheel drive vehicles. A guide is essential. Fortunately Morombe has an airport, and so the most convenient access to the centerline is by air.

The eclipse-seeker's eventual destination is the village of Ambahikily on the south side of the Mangoky River, where the road from Toliara to Morombe reaches its most northerly point, a few kilometers from the centerline. Though maps show a road leading northwest from Ambahikily across the middle of the track, no such route actually exists. Once off the highway there is only a direction to travel, not a route, for the trails are made mostly by bullock carts which travel freely in any direction without the confines of curb or ditch. Travel is easy, though slow, by four-wheel drive vehicles, but frequent stops are needed for directions from one community to another, and the vegetation is seldom open enough to allow a view of such a low altitude eclipse.

The Mangoky River is wide and steep-banked, (and inhabited by crocodiles), and most of the coastal beaches are blocked by inland swamps which harbor a magnificent collection of birds, but which make access nearly impossible. Morombe is likely the only reasonable location from which the eclipse could be viewed as it settles onto the Mozambique Channel, but this site will come with a severe time penalty, being about half-way toward the south limit.

Ambahikily is a bustling village about an hour-and-a-half from Morombe (37 km) with gregarious children who delight in having their pictures taken. Its market straddles the highway, presenting an opportunity to collect essential last-minute supplies for the eclipse. Open spaces are at a premium, but there are a number of sites in and around the village that will lend themselves to a good view of totality. We were unable to find any location that would allow the eclipse to be followed right to the horizon, and so those dedicated to seeing fourth contact will have to remain at the beach in Morombe.

Sunshine records for Toliara promise an average of 9.2 hours per day in June, 86% of the maximum possible, bettered only by the exceptionally dry conditions in Zambia and Angola. Some stations in western Madagascar north of the eclipse track (Maintirano, for instance) promise even more sunshine. This suggests that conditions along the centerline will be even more favorable than the statistics for Toliara would indicate. The fact that the eclipse comes late in June is also an advantage, for July is an even drier month on the island, and the statistics for June probably slightly overestimate the cloudiness that is characteristic of eclipse day.

At Morombe, average precipitation declines from 123 mm in January to 59 mm in March and to 6.8 mm in June. In recent years it has been only slightly variable, with no rain reported for June in 1997 and 9 mm reported in 1996. In 1972 a monthly rainfall of nearly 60 mm was recorded at Toliara, considerably larger than the normal 11 mm. These statistics indicate that the climatological record is a fairly reliable planning tool, though perhaps not so secure as over Zambia and Angola.

The prospects are not as good for the windward side of the island's mountain backbone. Taolagnaro, south of the eclipse track on the Indian coast, receives an average of 122 mm of precipitation in June, and the very muted dry season doesn't arrive until September. Cloudiness is common, with over 15% of observations at eclipse time reporting precipitation. High altitude locations southwest of Ihosy along the main highway will provide a very good view of the eclipse if the day is sunny, but the cloudiness is more of a threat here than along the west coast. The centerline crosses the route about half way between Ranohira and Ihosy, slightly on the leeward side of the mountain peaks, and so there is some help from the terrain in clearing the cloudiness which often lies along the east side. This is not a reliable location, but if forecasts prove promising or luck holds out, the view could be among the best along the entire track.

Satellite imagery for 1998 showed heavy cloud in the region north of Toliara on only 4 days of 37, comparable with the prospects near Lusaka. Southeastern Madagascar near Fort Dauphin, where the track is exposed to the uninterrupted flow of Indian Ocean winds, showed only seven days of heavy cloudiness, though there was often a band of sea breeze cloud inland from the coast. Watchers in this area will find that oceanfront locations are best, away from the sea breeze cloudiness a few kilometers inland.

It is difficult not to be pessimistic when examining the record for the windward side of Madagascar. The percent of possible sunshine has declined to 66% at Fort Dauphin and the frequency of broken or greater cloudiness is nearly 50%. This high frequency of broken cloudiness is further aggravated by the low sun angle, less than 10°, which further increases the chances of the eclipse view being blocked. With only a day needed to reach the much more suitable conditions in western Madagascar, it is likely that the southeast will not appeal to many travelers.

ON THE WATER

Should cruise boats attempt the eclipse in the waters of the Mozambique Channel, it appears from inspection of the satellite images from 1998 that the Madagascar side of the channel tends to have less cloud than the Mozambican side. On no occasion was the sky obscured from coast to coast, so a nautical eclipse venue should have a decent chance at sunny skies if forecasts can be obtained a day ahead in time to permit sailing to sunnier areas. Skies off of the Angolan coast tended to be slightly cloudier than off of Mozambique, but with numerous breaks that would permit easy relocation for a successful view of the Sun.

Wave heights average 1 to 1.5 meters on both the Angolan and Mozambique coasts, about half a meter lower than the average for Caribbean waters during the February 1998 eclipse. Sites close to the coast, particularly near Madagascar, should be able to find flatter seas.

SUMMARY

The first eclipse of the Millennium promises a bright start for the next thousand years with several excellent observation sites. Because of the Angolan civil war, the prime locations will be in western Zambia, with excellent access to the track. Second place must go to southwest Madagascar, protected by the mountain barrier that divides the island, though travel to the centerline there is more difficult. Zimbabwe, especially in the west, and Zambia in the east, offer good cloud prospects and a well-developed infrastructure, while Mozambique suffers from a high level of cloudiness in the Zambezi River lowlands.

WEATHER WEB SITES

1. http://www.tvweather.com
     A good starting point. This site has many links to current and past weather around the world.
2. http://satpix.nottingham.ac.uk/ - Good site for weather Satellite images of Africa.
3. http://www.zamnet.zm/zamnet/zmd/zmd1.htm - Zambian Weather Office (still under construction).
4. http://weather.utande.co.zw/Office/index.htm - Zimbabwe Weather Office.
5. http://cirrus.sawb.gov.za/ - South Africa Weather Bureau.
6. http://www.worldclimate.com/climate/index.htm - World data base of temperature and rainfall.

Observing the Eclipse

Eye Safety and Solar Eclipses

B. Ralph Chou, MSc, OD
Associate Professor, School of Optometry, University of Waterloo
Waterloo, Ontario, Canada N2L 3G1

A total solar eclipse is probably the most spectacular astronomical event that most people will experience in their lives. There is a great deal of interest in watching eclipses, and thousands of astronomers (both amateur and professional) travel around the world to observe and photograph them.

A solar eclipse offers students a unique opportunity to see a natural phenomenon that illustrates the basic principles of mathematics and science that are taught through elementary and secondary school. Indeed, many scientists (including astronomers!) have been inspired to study science as a result of seeing a total solar eclipse. Teachers can use eclipses to show how the laws of motion and the mathematics of orbital motion can predict the occurrence of eclipses. The use of pinhole cameras and telescopes or binoculars to observe an eclipse leads to an understanding of the optics of these devices. The rise and fall of environmental light levels during an eclipse illustrate the principles of radiometry and photometry, while biology classes can observe the associated behavior of plants and animals. It is also an opportunity for children of school age to contribute actively to scientific research - observations of contact timings at different locations along the eclipse path are useful in refining our knowledge of the orbital motions of the moon and earth, and sketches and photographs of the solar corona can be used to build a three-dimensional picture of the sun's extended atmosphere during the eclipse.

However, observing the Sun can be dangerous if you do not take the proper precautions. The solar radiation that reaches the surface of the earth includes ultraviolet (UV) radiation at wavelengths longer than 290 nm, to radio waves in the meter range. The tissues in the eye transmit a substantial part of the radiation between 380 and 1400 nm to the light-sensitive retina at the back of the eye. While environmental exposure to UV radiation is known to contribute to the accelerated aging of the outer layers of the eye and the development of cataracts, the concern over improper viewing of the Sun during an eclipse is for the development of "eclipse blindness" or retinal burns.

Exposure of the retina to intense visible light causes damage to its light-sensitive rod and cone cells. The light triggers a series of complex chemical reactions within the cells which damages their ability to respond to a visual stimulus, and in extreme cases, can destroy them. The result is a loss of visual function which may be either temporary or permanent, depending on the severity of the damage. When a person looks repeatedly or for a long time at the Sun without proper protection for the eyes, this photochemical retinal damage may be accompanied by a thermal injury - the high level of visible and near-infrared radiation causes heating that literally cooks the exposed tissue. This thermal injury or photocoagulation destroys the rods and cones, creating a small blind area. The danger to vision is significant because photic retinal injuries occur without any feeling of pain (the retina has no pain receptors), and the visual effects do not occur for at least several hours after the damage is done. (Pitts, 1993)

The only time that the Sun can be viewed safely with the naked eye is during a total eclipse, when the moon completely covers the Sun. *It is never safe to look at a partial or annular eclipse, or the partial phases of a total eclipse, without the proper equipment and techniques.* Even when 99% of the sun's surface (the photosphere) is obscured during the partial phases of a solar eclipse, the remaining crescent Sun is still intense enough to cause a retinal burn, even though illumination levels are comparable to twilight (Chou, 1981, 1996; Marsh, 1982). Failure to use proper observing methods may result in permanent eye damage or severe visual loss. This can have important adverse effects on career choices and earning potential, since it has been shown that most individuals who sustain eclipse-related eye injuries are children and young adults (Penner and McNair, 1966; Chou and Krailo, 1981).

The same techniques for observing the Sun outside of eclipses are used to view and photograph annular solar eclipses and the partly eclipsed Sun (Sherrod, 1981; Pasachoff & Menzel 1992; Pasachoff & Covington, 1993; Reynolds & Sweetsir, 1995). The safest and most inexpensive method is by projection. A pinhole or small opening is used to form an image of the Sun on a screen placed about a meter behind the opening. Multiple openings in perfboard, a loosely woven straw hat, or even between interlaced fingers can be used to cast a pattern of solar images on a screen. A similar effect is seen on the ground below a broad-leafed tree: the many "pinholes" formed by overlapping leaves creates hundreds of crescent-shaped images. Binoculars or a small telescope mounted on a tripod can also be used to project a magnified image of the Sun onto a white card. All of these methods can be used to provide a safe view of the partial phases of an eclipse to a group of observers, but care must be taken to ensure that no-one looks through the device. The main advantage of the projection methods is that nobody is looking directly at the Sun. The disadvantage of the pinhole method is that the screen must be placed at least a meter behind the opening to get a solar image that is large enough to see easily.

The Sun can only be viewed directly when filters specially designed to protect the eyes are used. Most of these filters have a thin layer of chromium alloy or aluminum deposited on their surfaces that attenuates both visible and near-infrared radiation. A safe solar filter should transmit less than 0.003% (density~4.5)[10] of visible light (380 to 780 nm) and no more than 0.5% (density~2.3) of the near-infrared radiation (780 to 1400 nm). Figure 14 shows transmittance curves for a selection of safe solar filters.

One of the most widely available filters for safe solar viewing is shade number 14 welder's glass, which can be obtained from welding supply outlets. A popular inexpensive alternative is aluminized polyester[11] that has been made specially for solar observation. ("Space blankets" and aluminized polyester used in gardening are NOT suitable for this purpose!) Unlike the welding glass, aluminized polyester can be cut to fit any viewing device, and doesn't break when dropped. It has recently been pointed out that some aluminized polyester filters may have large (up to approximately 1 mm in size) defects in their aluminum coatings that may be hazardous. A microscopic analysis of examples of such defects shows that despite their appearance, the defects arise from a hole in one of the two aluminized polyester films used in the filter. There is no large opening completely devoid of the protective aluminum coating. While this is a quality control problem, the presence of a defect in the aluminum coating does not necessarily imply that the filter is hazardous. When in doubt, an aluminized polyester solar filter that has coating defects larger than 0.2 mm in size, or more than a single defect in any 5 mm circular zone of the filter, should not be used.

An alternative to aluminized polyester solar filter material that has become quite popular is "black polymer" in which carbon particles are suspended in a resin matrix. This material is somewhat stiffer than aluminized polyester and requires a special holding cell if it is to be used at the front of binoculars, telephoto lenses or telescopes. Intended mainly as a visual filter, the polymer gives a yellow image of the Sun (aluminized polyester produces a blue-white image). This type of filter may show significant variations in density of the tint across its extent; some areas may appear much lighter than others. Lighter areas of the filter transmit more infrared radiation than may be desirable. A recent development is a filter that consists of aluminum-coated black polymer. Combining the best features of polyester and black polymer, this new material may eventually replace both as the filter of choice in solar eclipse viewers. The transmittance curve of one form of this hybrid filter (Polymer Plus™ by Thousand Oaks Optical) is shown in Figure 14.

Many experienced solar observers use one or two layers of black-and-white film that has been fully exposed to light and developed to maximum density. The metallic silver contained in the film emulsion is the protective filter, however any black-and-white negative with images in it is not suitable for this purpose. More recently, solar observers have used floppy disks and compact disks (CDs and CD-ROMs) as protective filters by covering the central openings and looking through the disk media. However, the optical quality of the solar image formed by a floppy disk or CD is relatively poor compared to aluminized polyester or welder's glass. Some CDs are made with very thin aluminum coatings which are not safe - if you can see through the CD in normal room lighting, don't use it!! No filter should be used with an optical device (e.g. binoculars, telescope, camera) unless it has been specifically designed for that purpose and is mounted at the front end. Some sources of solar filters are listed below.

Unsafe filters include color film, black-and-white film that contains no silver, film negatives with images on them, smoked glass, sunglasses (single or multiple pairs), photographic neutral density filters and polarizing filters. Most of these transmit high levels of invisible infrared radiation which can cause a thermal retinal burn (see Figure 14). The fact that the Sun appears dim, or that you feel no discomfort when looking at the Sun through the filter, is no guarantee that your eyes are safe. Solar filters designed to thread into eyepieces that are often provided with inexpensive telescopes are also unsafe. These glass filters often crack unexpectedly from overheating when the telescope is pointed at the Sun, and retinal damage can occur faster than the observer can move the eye from the eyepiece. Avoid unnecessary risks. Your local planetarium, science center, or amateur astronomy club can provide additional information on how to observe the eclipse safely.

---

[10] In addition to the term transmittance (in percent), the energy transmission of a filter can also be described by the term density (unitless) where density 'd' is the common logarithm of the reciprocal of transmittance 't' or $d = \log_{10}[1/t]$. A density of '0' corresponds to a transmittance of 100%; a density of '1' corresponds to a transmittance of 10%; a density of '2' corresponds to a transmittance of 1%, etc....

[11] Aluminized polyester is popularly known to as mylar. DuPont actually owns the trademark "Mylar™" and does not manufacture this material for use as a solar filter.

There are some concerns that UVA radiation (wavelengths between 315 and 380 nm) in sunlight may also adversely affect the retina (Del Priore, 1991). While there is some experimental evidence for this, it only applies to the special case of aphakia, where the natural lens of the eye has been removed because of cataract or injury, and no UV-blocking spectacle, contact or intraocular lens has been fitted. In an intact normal human eye, UVA radiation does not reach the retina because it is absorbed by the crystalline lens. In aphakia, normal environmental exposure to solar UV radiation may indeed cause chronic retinal damage. However, the solar filter materials discussed in this article attenuate solar UV radiation to a level well below the minimum permissible occupational exposure for UVA (ACGIH, 1994), so an aphakic observer is at no additional risk of retinal damage when looking at the Sun through a proper solar filter.

In the days and weeks before a solar eclipse occurs, there are often news stories and announcements in the media, warning about the dangers of looking at the eclipse. Unfortunately, despite the good intentions behind these messages, they frequently contain misinformation, and may be designed to scare people from seeing the eclipse at all. However, this tactic may backfire, particularly when the messages are intended for students. A student who heeds warnings from teachers and other authorities not to view the eclipse because of the danger to vision, and learns later that other students did see it safely, may feel cheated out of the experience. Having now learned that the authority figure was wrong on one occasion, how is this student going to react when other health-related advice about drugs, AIDS, or smoking is given? Misinformation may be just as bad, if not worse than no information.

Remember that the *total* phase of an eclipse can and should be seen without any filters, and certainly never by projection! It is completely safe to do so. Even after observing 14 solar eclipses, I find the naked eye view of the totally eclipsed Sun awe-inspiring. I hope you will also enjoy the experience.

SOURCES FOR SOLAR FILTERS

The following is a brief list of sources for aluminized polyester and/or glass filters specifically designed for safe solar viewing with or without a telescope. The list is not meant to be exhaustive, but is simply a representative sample of sources for solar filters currently available in North America and Europe. For additional sources, see advertisements in *Astronomy* and/or *Sky & Telescope* magazines. The inclusion of any source on this list does not imply an endorsement of that source by the authors or NASA.

• ABELexpress - Astronomy Division, 100 Rosslyn Rd., Carnegie, PA 15106. (412) 279-0672
• American Paper Optics, 3080 Bartlett Corporate Drive, Bartlett, TN 38133. (800)767-8427
• Celestron International, 2835 Columbia St., Torrance, CA 90503. (310) 328-9560
• Edwin Hirsch, 29 Lakeview Dr., Tomkins Cove, NY 10986. (914) 786-3738
• Meade Instruments Corporation, 16542 Millikan Ave., Irvine, CA 92714. (714) 756-2291
• Orion Telescopes and Binoculars, P.O. Box 1815, Santa Cruz, CA 95061-1815. (800) 447-1001
• Pocono Mountain Optics, 104 NP 502 Plaza, Moscow, PA 18444. (717) 842-1500
• Rainbow Symphony, Inc., 6860 Canby Ave., #120, Reseda, CA 91335 (800) 821-5122
• Roger W. Tuthill, Inc., 11 Tanglewood Lane, Mountainside, NJ 07092. (908) 232-1786
• Thousand Oaks Optical, Box 5044-289, Thousand Oaks, CA 91359. (805) 491-3642
• Khan Scope Centre, 3243 Dufferin Street, Toronto, Ontario, Canada M6A 2T2 (416) 783-4140
• Perceptor Telescopes TransCanada, Brownsville Junction Plaza, Box 38,
    Schomberg, Ontario, Canada L0G 1T0 (905) 939-2313
• Swan Packaging Ltd., Unit 6, Princewood Road, Earlstrees Industrial Estate, Corby,
    Northants NN17 4AP United Kingdom. +44(01536)204272
• Eclipse 99 Ltd., Belle Etoile, Rue du Hamel, Guernsey GY5 7QJ. 001 44 1481 64847

IAU SOLAR ECLIPSE EDUCATION COMMITTEE

In order to ensure that astronomers and public health authorities have access to information related to safe viewing practices, the International Astronomical Union, the international organization for professional astronomers, set up a Solar Eclipse Education Committee. Under Prof. Jay M. Pasachoff of Williams College, the Committee has assembled information on safe methods of observing the Sun and solar eclipses, eclipse-related eye injuries, and samples of educational materials on solar eclipses.

For more information, contact Prof. Jay M. Pasachoff, Hopkins Observatory, Williams College, Williamstown, MA 01267, USA (e-mail: jay.m.pasachoff@williams.edu). Information on safe solar filters can be obtained by contacting Dr. B. Ralph Chou (e-mail: bchou@sciborg.uwaterloo.ca).

ECLIPSE PHOTOGRAPHY

The eclipse may be safely photographed provided that the above precautions are followed. Almost any kind of camera with manual controls can be used to capture this rare event. However, a lens with a fairly long focal length is recommended to produce as large an image of the Sun as possible. A standard 50 mm lens yields a minuscule 0.5 mm image, while a 200 mm telephoto or zoom produces a 1.9 mm image. A better choice would be one of the small, compact catadioptic or mirror lenses that have become widely available in the past ten years. The focal length of 500 mm is most common among such mirror lenses and yields a solar image of 4.6 mm. With one solar radius of corona on either side, an eclipse view during totality will cover 9.2 mm. Adding a 2x tele-converter will produce a 1000 mm focal length, which doubles the Sun's size to 9.2 mm. Focal lengths in excess of 1000 mm usually fall within the realm of amateur telescopes. If full disk photography of partial phases on 35 mm format is planned, the focal length of the optics must not exceed 2600 mm. However, since most cameras don't show the full extent of the image in their viewfinders, a more practical limit is about 2000 mm. Longer focal lengths permit photography of only a magnified portion of the Sun's disk. In order to photograph the Sun's corona during totality, the focal length should be no longer than 1500 mm to 1800 mm (for 35 mm equipment). However, a focal length of 1000 mm requires less critical framing and can capture some of the longer coronal streamers. For any particular focal length, the diameter of the Sun's image is approximately equal to the focal length divided by 109 (Table 18).

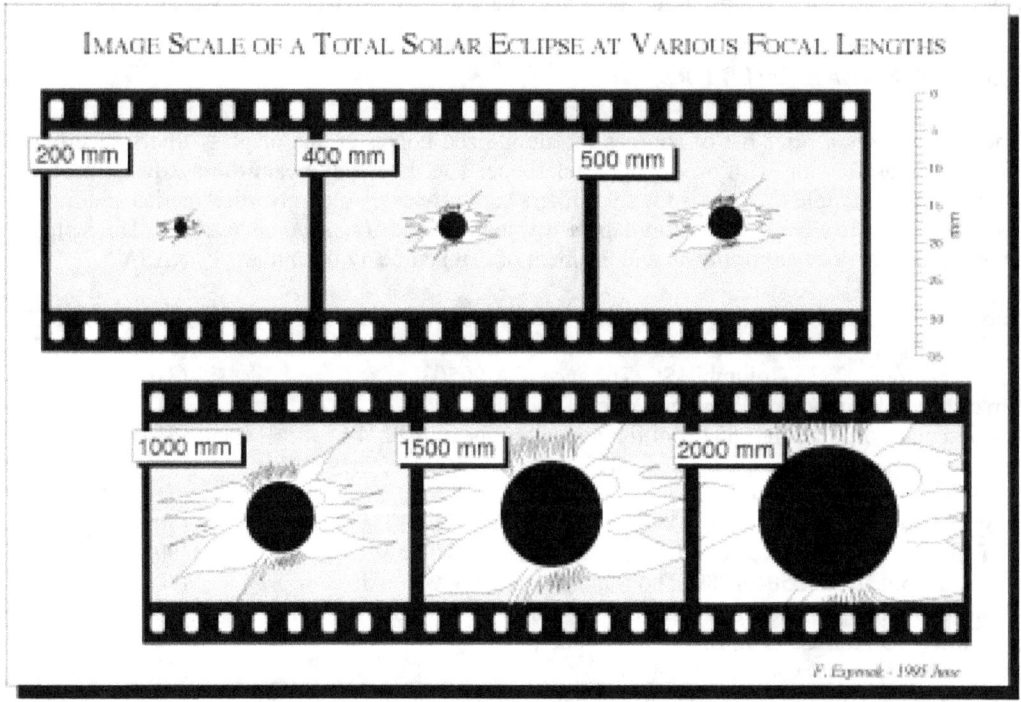

A solar filter must be used on the lens throughout the partial phases for both photography and safe viewing. Such filters are most easily obtained through manufacturers and dealers listed in *Sky & Telescope* and *Astronomy* magazines (see: SOURCES FOR SOLAR FILTERS). These filters typically attenuate the Sun's visible and infrared energy by a factor of 100,000. However, the actual filter factor and choice of ISO film speed will play critical roles in determining the correct photographic exposure. A low to medium speed film is recommended (ISO 50 to 100) since the Sun gives off abundant light. The easiest method for determining the correct exposure is accomplished by running a calibration test on the uneclipsed Sun. Shoot a roll of film of the mid-day Sun at a fixed aperture (f/8 to f/16) using every shutter speed between 1/1000 and 1/4 second. After the film is developed, note the best exposures and use them to photograph all the partial phases. The Sun's surface brightness remains constant throughout the eclipse, so no exposure compensation is needed except for the crescent phases which require two more stops due to solar limb darkening. Bracketing by several stops is also necessary if haze or clouds interfere on eclipse day.

Certainly the most spectacular and awe inspiring phase of the eclipse is totality. For a few brief minutes or seconds, the Sun's pearly white corona, red prominences and chromosphere are visible. The great challenge is to obtain a set of photographs which captures some aspect of these fleeting phenomena. The most important point to remember is that during the total phase, all solar filters *must be removed!* The corona has a surface brightness a million times fainter than the photosphere, so photographs of the corona are made without a filter. Furthermore, it is completely safe to view the totally eclipsed Sun directly with the naked eye. No filters are needed and they will only hinder your view. The average brightness of the corona varies inversely with the distance from the Sun's limb. The inner corona is far brighter than the outer corona. Thus, no single exposure can capture its full dynamic range. The best strategy is to choose one aperture or f/number and bracket the exposures over a range of shutter speeds (i.e., 1/1000 down to 1 second). Rehearsing this sequence is highly recommended since great excitement accompanies totality and there is little time to think.

Exposure times for various combinations of film speeds (ISO), apertures (f/number) and solar features (chromosphere, prominences, inner, middle and outer corona) are summarized in Table 19. The table was developed from eclipse photographs made by Espenak as well as from photographs published in *Sky and Telescope*. To use the table, first select the ISO film speed in the upper left column. Next, move to the right to the desired aperture or f/number for the chosen ISO. The shutter speeds in that column may be used as starting points for photographing various features and phenomena tabulated in the 'Subject' column at the far left. For example, to photograph prominences using ISO 100 at f/11, the table recommends an exposure of 1/500. Alternatively, you can calculate the recommended shutter speed using the 'Q' factors tabulated along with the exposure formula at the bottom of Table 19. Keep in mind that these exposures are based on a clear sky and a corona of average brightness. You should bracket your exposures one or more stops to take into account the actual sky conditions and the variable nature of these phenomena.

Another interesting way to photograph the eclipse is to record its phases all on one frame. This is accomplished by using a stationary camera capable of making multiple exposures (check the camera instruction manual). Since the Sun moves through the sky at the rate of 15 degrees per hour, it slowly drifts through the field of view of any camera equipped with a normal focal length lens (i.e., 35 to 50 mm). If the camera is oriented so that the Sun drifts along the frame's diagonal, it will take over three hours for the Sun to cross the field of a 50 mm lens. The proper camera orientation can be determined through trial and error several days before the eclipse. This will also insure that no trees or buildings obscure the view during the eclipse. The Sun should be positioned along the eastern (left in the northern hemisphere) edge or corner of the viewfinder shortly before the eclipse begins. Exposures are then made throughout the eclipse at ~five minute intervals. The camera must remain perfectly rigid during this period and may be clamped to a wall or post since tripods are easily bumped. If you're in the path of totality, remove the solar filter during the total phase and take a long exposure (~1 second) in order to record the corona in your sequence. The final photograph will consist of a string of Suns, each showing a different phase of the eclipse.

Finally, an eclipse effect that is easily captured with point-and-shoot or automatic cameras should not be overlooked. Use a kitchen sieve or colander and allow its shadow to fall on a piece of white cardboard placed several feet away. The holes in the utensil act like pinhole cameras and each one projects its own image of the Sun. The effect can also be duplicated by forming a small aperture with one's hands and watching the ground below. The pinhole camera effect becomes more prominent with increasing eclipse magnitude. Virtually any camera can be used to photograph the phenomenon, but automatic cameras must have their flashes turned off since this would otherwise obliterate the pinhole images.

Several comments apply to those who choose to photograph the eclipse aboard a cruise ship at sea. Shipboard photography puts certain limits on the focal length and shutter speeds that can be used. It's difficult to make specific recommendations since it depends on the stability of the ship as well as wave heights encountered on eclipse day. Certainly telescopes with focal lengths of 1000 mm or more can be ruled out since their small fields of view would require the ship to remain virtually motionless during totality, and this is rather unlikely even given calm seas. A 500 mm lens might be a safe upper limit in focal length. Film choice could be determined on eclipse day by viewing the Sun through the camera lens and noting the image motion due to the rolling sea. If it's a calm day, you might try an ISO 100 film. For rougher seas, ISO 400 or more might be a better choice. Shutter speeds as slow as 1/8 or 1/4 may be tried if the conditions warrant it. Otherwise, stick with a 1/15 or 1/30 and shoot a sequence through 1/1000 second. It might be good insurance to bring a wider 200 mm lens just in case the seas are rougher than expected. As worst case scenario, Espenak photographed the 1984 total eclipse aboard a 95 foot yacht in seas of 3 feet. He had to hold on with one hand and point his 350 mm lens with the other! Even at that short focal length, it was difficult to keep the Sun in the field. However, any large cruise ship will offer a far more stable platform than this.

For more information on eclipse photography, observations and eye safety, see FURTHER READING in the BIBLIOGRAPHY.

SKY AT TOTALITY

The total phase of an eclipse is accompanied by the onset of a rapidly darkening sky whose appearance resembles evening twilight about 30 to 40 minutes after sunset. The effect presents an excellent opportunity to view planets and bright stars in the daytime sky. Aside from the sheer novelty of it, such observations are useful in gauging the apparent sky brightness and transparency during totality.

During the total solar eclipse of 2001, the Sun is in Gemini. Depending on the geographic location, three or four naked eye planets as well as a number of bright stars will be above the horizon within the umbral path. Figure 15 depicts the appearance of the sky during totality as seen from the center line at 13:00 UT. This corresponds to western Zambia near the Angolan border.

The most conspicuous planet visible will be Jupiter (m_V=−1.5) just 5° west of the Sun. Although Mercury (m_V=+2.7) is 8.6° west of the Sun, it is quite faint and will be difficult to spot. With a modest phase of 0.03, Mercury's illuminated face points away from Earth since the planet passed through inferior conjunction just five days earlier. Saturn (m_V=+0.3) should prove an easier target 22.6° west of the Sun. Venus (m_V=−3.3) is brightest of all, but the planet is near its greatest elongation 45.3° west of the Sun. Venus will be low in the west for observers in Angola and western Zambia. East of those positions, the planet will have already set before totality begins. Mars is near opposition and is below the horizon for all points along the entire umbral path.

A number of the brightest winter stars may also be visible during totality. Capella (m_V=+0.08) is 24° north of the Sun while Castor (m_V=+1.94) and Pollux (m_V=+1.14) stand 22° and 24° to the east. Procyon (m_V=+0.38) and Sirius (m_V=−1.46) are located 30° and 42° to the southeast, respectively. Betelgeuse (m_V=+0.5v) and Rigel (m_V=+0.12) are south at 16° and 34°, while Aldebaran (m_V=+0.85) is 21° to the west. Finally, Canopus (m_V=−0.72) lies high in the sky 76° due south. Star visibility requires a very dark and cloud free sky during totality.

The following ephemeris [using Bretagnon and Simon, 1986] gives the positions of the naked eye planets during the eclipse. *Delta* is the distance of the planet from Earth (A.U.'s), *App. Mag.* is the apparent visual magnitude of the planet, and *Solar Elong* gives the elongation or angle between the Sun and planet.

```
Ephemeris: 2001 Jun  21  13:00:00 UT                    Equinox = Mean Date
```

Planet	RA	Declination	Delta	App. Mag.	Apparent Diameter "	Phase	Solar Elong °
Sun	06h00m56s	+23°26'19"	1.01628	-26.7	1888.6	-	-
Moon	06h03m27s	+22°55'26"	0.00245	12.2	1955.9	-0.00	0.8E
Mercury	05h29m52s	+18°53'11"	0.57273	2.7	11.7	0.03	8.6W
Venus	02h53m30s	+13°40'18"	0.80823	-3.8	20.6	0.56	45.3W
Mars	17h16m20s	-26°44'01"	0.45018	-2.1	20.8	1.00	169.4E
Jupiter	05h38m52s	+23°02'33"	6.11294	-1.5	32.2	1.00	5.1W
Saturn	04h25m15s	+19°56'56"	10.01774	0.3	16.6	1.00	22.5W

For sky maps from other locations along the path of totality, see the special web site for the total solar eclipse of 2001: *http://sunearth.gsfc.nasa.gov/eclipse/TSE2001/TSE2001.html*

CONTACT TIMINGS FROM THE PATH LIMITS

Precise timings of beading phenomena made near the northern and southern limits of the umbral path (i.e., the graze zones), are of value in determining the diameter of the Sun relative to the Moon at the time of the eclipse. Such measurements are essential to an ongoing project to monitor changes in the solar diameter. Due to the conspicuous nature of the eclipse phenomena and their strong dependence on geographical location, scientifically useful observations can be made with relatively modest equipment. A small telescope, short wave radio and portable camcorder are usually used to make such measurements. Time signals are broadcast via short wave stations WWV and CHU, and are recorded simultaneously as the eclipse is videotaped. If a video camera is not available, a tape recorder can be used to record time signals with verbal timings of each event. Inexperienced observers are cautioned to use great care in making such observations. The safest timing technique consists of observing a projection of the Sun rather than directly imaging the solar disk itself. The observer's geodetic coordinates are required and can be measured from USGS or other large scale maps. If a map is unavailable, then a detailed description of the observing site should be included which provides information such as distance and directions of the nearest towns/settlements, nearby landmarks, identifiable buildings and road intersections. The method of contact timing should be described in detail, along with an estimate of the error. The precisional requirements of these observations are ±0.5 seconds in time, 1" (~30 meters) in latitude and longitude, and ±20 meters (~60 feet) in elevation. Although GPS's (Global Positioning Satellite receivers) are commercially available (~$150 US), their positional accuracy of ±100 meters is about three times larger than the minimum accuracy required by grazing eclipse measurements. GPS receivers are also a useful source for accurate UT. The International Occultation Timing Association (IOTA) coordinates observers world-wide during each eclipse. For more information, contact:

Dr. David W. Dunham, IOTA   Web Site: http://www.lunar-occultations.com/iota
7006 Megan Lane   E-mail: dunham@erols.com
Greenbelt, MD 20770-3012, USA   Phone: (301) 474-4722

Send reports containing graze observations, eclipse contact and Baily's bead timings, including those made anywhere near or in the path of totality or annularity to:

Dr. Alan D. Fiala
Orbital Mechanics Dept.
U. S. Naval Observatory
3450 Massachusetts Ave., NW
Washington, DC 20392-5420, USA

PLOTTING THE PATH ON MAPS

If high resolution maps of the umbral path are needed, the coordinates listed in Tables 7 and 8 are conveniently provided in longitude increments of 1° and 30' respectively to assist plotting by hand. The path coordinates in Table 3 define a line of maximum eclipse at five minute increments in Universal Time. If observations are to be made near the limits, then the grazing eclipse zones tabulated in Table 8 should be used. A higher resolution table of graze zone coordinates at longitude increments of 7.5' is available via a special web site for the 2001 total eclipse (*http://sunearth.gsfc.nasa.gov/eclipse/TSE2001/TSE2001.html*). Global Navigation Charts (1:5,000,000), Operational Navigation Charts (scale 1:1,000,000) and Tactical Pilotage Charts (1:500,000) of many parts of the world are published by the National Imagery and Mapping Agency (formerly known as the Defense Mapping Agency). Sales and distribution of these maps is through the National Ocean Service (NOS). For specific information about map availability, purchase prices, and ordering instructions, contact the NOS at:

NOAA Distribution Division, N/ACC3   phone: 301-436-8301
National Ocean Service   FAX: 301-436-6829
Riverdale, MD 20737-1199, USA

It is also advisable to check the telephone directory for any map specialty stores in your city or metropolitan area. They often have large inventories of many maps available for immediate delivery.

IAU WORKING GROUP ON ECLIPSES

Professional scientists are asked to send descriptions of their eclipse plans to the Working Group on Eclipses of the International Astronomical Union, so that they can keep a list of observations planned. Send such descriptions, even in preliminary form, to:

International Astronomical Union/Working Group on Eclipses
Prof. Jay M. Pasachoff, Chair
Williams College–Hopkins Observatory   email: jay.m.pasachoff@williams.edu
Williamstown, MA 01267, USA   FAX: (413) 597-3200

The members of the Working Group on Eclipses of Commissions 10 and 12 of the International Astronomical Union are: Jay M. Pasachoff (USA), Chair; F. Clette (Belgium), F. Espenak (USA); Iraida Kim (Russia); V. Rusin (Slovakia); Jagdev Singh (India); M. Stavinschi (Romania); Yoshinori Suematsu (Japan); consultant: J. Anderson (Canada).

ECLIPSE DATA ON INTERNET

NASA ECLIPSE BULLETINS ON INTERNET

To make the NASA solar eclipse bulletins accessible to as large an audience as possible, these publications are also available via the Internet. This was made possible through the efforts and expertise of Dr. Joe Gurman (GSFC/Solar Physics Branch). All future eclipse bulletins will be available via Internet.

NASA eclipse bulletins can be read or downloaded via the World-Wide Web using a Web browser (e.g.: Netscape, Microsoft Explorer, etc.) from the GSFC SDAC (Solar Data Analysis Center) Eclipse Information home page, or from top-level URL's for the currently available eclipse bulletins themselves:

http://umbra.nascom.nasa.gov/eclipse/   (SDAC Eclipse Information)

http://umbra.nascom.nasa.gov/eclipse/941103/rp.html   (1994 Nov 3)
http://umbra.nascom.nasa.gov/eclipse/951024/rp.html   (1995 Oct 24)
http://umbra.nascom.nasa.gov/eclipse/970309/rp.html   (1997 Mar 9)
http://umbra.nascom.nasa.gov/eclipse/980226/rp.html   (1998 Feb 26)
http://umbra.nascom.nasa.gov/eclipse/990811/rp.html   (1999 Aug 11)
http://umbra.nascom.nasa.gov/eclipse/000621/rp.html   (2001 Jun 21)

The original Microsoft Word text files, GIF and PICT figures (Macintosh format) are also available via anonymous ftp. They are stored as BinHex-encoded, StuffIt-compressed Mac folders with .hqx suffixes. For PC's, the text is available in a zip-compressed format in files with the .zip suffix. There are three sub directories for figures (GIF format), maps (JPEG format), and tables (html tables, easily readable as plain text). For example, NASA RP 1383 (Total Solar Eclipse of 1998 February 26 [=980226]) has a directory for these files is as follows:

file://umbra.nascom.nasa.gov/pub/eclipse/980226/RP1383GIFs.hqx
file://umbra.nascom.nasa.gov/pub/eclipse/980226/RP1383PICTs.hqx
file://umbra.nascom.nasa.gov/pub/eclipse/980226/RP1383text.hqx
file://umbra.nascom.nasa.gov/pub/eclipse/980226/RP1383text.zip
file://umbra.nascom.nasa.gov/pub/eclipse/980226/figures   (directory with GIF's)
file://umbra.nascom.nasa.gov/pub/eclipse/980226/maps   (directory with JPEG's)
file://umbra.nascom.nasa.gov/pub/eclipse/980226/tables   (directory with html's)

Other eclipse bulletins have a similar directory format.

Current plans call for making all future NASA eclipse bulletins available over the Internet, at or before publication of each. The primary goal is to make the bulletins available to as large an audience as possible. Thus, some figures or maps may not be at their optimum resolution or format. Comments and suggestions are actively solicited to fix problems and improve on compatibility and formats.

FUTURE ECLIPSE PATHS ON INTERNET

Presently, the NASA eclipse bulletins are published 24 to 36 months before each eclipse. However, there have been a growing number of requests for eclipse path data with an even greater lead time. To accommodate the demand, predictions have been generated for all central solar eclipses from 1991 through 2030. All predictions are based on j=2 ephemerides for the Sun [Newcomb, 1895] and Moon [Brown, 1919, and Eckert, Jones and Clark, 1954]. The value used for the Moon's secular acceleration is n-dot = -26 arc-sec/cy*cy, as deduced by Morrison and Ward [1975]. A correction of -0.6" was added to the Moon's ecliptic latitude to account for the difference between the Moon's center of mass and center of figure. The value for delta-T is from direct measurements during the 20th century and extrapolation into the 21st century. The value used for the Moon's mean radius is k=0.272281.

The umbral path characteristics have been predicted at 2 minute intervals of time compared to the 6 minute interval used in *Fifty Year Canon of Solar Eclipses: 1986-2035* [Espenak, 1987]. This should provide enough detail for making preliminary plots of the path on larger scale maps. Global maps using an orthographic projection also present the regions of partial and total (or annular) eclipse. The index page for the path tables and maps is:

http://sunearth.gsfc.nasa.gov/eclipse/SEpath/SEpath.html

SPECIAL WEB SITE FOR 2001 SOLAR ECLIPSE

A special web site is being set up to supplement this bulletin with additional predictions, tables and data for the total solar eclipse of 2001. Some of the data posted there include an expanded version of Table 8 (Mapping Coordinates for the Zones of Grazing Eclipse), and local circumstance tables with additional cities as well as for astronomical observatories. Also featured will be higher resolution maps of selected sections of the path of totality and limb profile figures for a range of locations/times along the path. The URL of the special TSE2001 site is:

http://sunearth.gsfc.nasa.gov/eclipse/TSE2001/TSE2001.html

ANNULAR SOLAR ECLIPSES OF 2001 AND 2002

Following the total solar eclipse of 2001, two annular eclipses will occur on 2001 December 14 and 2002 June 10. Both of these events take place in the Western Hemisphere with the paths of annularity confined primarily to the Pacific Ocean. By coincidence, the eastern end of each path terminates in Central America (Figure 16). No NASA eclipse bulletins are planned for these events. However, additional predictions, maps and tables will be made available on the web:

http://sunearth.gsfc.nasa.gov/eclipse/eclipse.html

TOTAL SOLAR ECLIPSE OF 2002 DEC 04

The next total eclipse of the Sun is also visible from southern Africa. The path of the Moon's umbral shadow begins in the South Atlantic, off the west coast of equatorial Africa. It crosses through Angola, Zambia, Namibia, Botswana, Zimbabwe, South Africa and Mozambique (Figure 17). Totality takes place in the morning hours with a central duration ranging from 1 to 1.5 minutes. The track continues across the Indian Ocean and ends in southern Australia north of Adelaide (Figure 18).

Complete details will eventually be posted on the NASA TSE2002 web site as well as in the next NASA bulletin scheduled for publication in late 2000. The TSE2002 web site address is:

http://sunearth.gsfc.nasa.gov/eclipse/TSE2002/TSE2002.html

PREDICTIONS FOR ECLIPSE EXPERIMENTS

This publication provides comprehensive information on the 2001 total solar eclipse to both the professional and amateur/lay communities. However, certain investigations and eclipse experiments may require additional information which lies beyond the scope of this work. We invite the international professional community to contact us for assistance with any aspect of eclipse prediction including predictions for locations not included in this publication, or for more detailed predictions for a specific location (e.g.: lunar limb profile and limb corrected contact times for an observing site).

This service is offered for the 2001 eclipse as well as for previous eclipses in which analysis is still in progress. To discuss your needs and requirements, please contact Fred Espenak (espenak@gsfc.nasa.gov).

ALGORITHMS, EPHEMERIDES AND PARAMETERS

Algorithms for the eclipse predictions were developed by Espenak primarily from the *Explanatory Supplement* [1974] with additional algorithms from Meeus, Grosjean and Vanderleen [1966] and Meeus [1982]. The solar and lunar ephemerides were generated from the JPL DE200 and LE200, respectively. All eclipse calculations were made using a value for the Moon's radius of $k=0.2722810$ for umbral contacts, and $k=0.2725076$ (adopted IAU value) for penumbral contacts. Center of mass coordinates were used except where noted. Extrapolating from 1997 to 1999, a value for ΔT of 65.0 seconds was used to convert the predictions from Terrestrial Dynamical Time to Universal Time. The international convention of presenting date and time in descending order has been used throughout the bulletin (i.e., *year, month, day, hour, minute, second*).

The primary source for geographic coordinates used in the local circumstances tables is *The New International Atlas* (Rand McNally, 1991). Elevations for major cities were taken from *Climates of the World* (U. S. Dept. of Commerce, 1972).

All eclipse predictions presented in this publication were generated on a Macintosh PowerPC 8500 computer. Word processing and page layout for the publication were done using Microsoft Word v5.1. Figures were annotated with Claris MacDraw Pro 1.5. Meteorological diagrams were prepared using Corel Draw 5.0 and converted to Macintosh compatible files. Finally, the bulletin was printed on a 600 dpi laser printer (Apple LaserWriter Pro).

The names and spellings of countries, cities and other geopolitical regions are not authoritative, nor do they imply any official recognition in status. Corrections to names, geographic coordinates and elevations are actively solicited in order to update the data base for future eclipses. All calculations, diagrams and opinions presented in this publication are those of the authors and they assume full responsibility for their accuracy.

BIBLIOGRAPHY

REFERENCES

Bretagnon, P., and Simon, J. L., *Planetary Programs and Tables from –4000 to +2800*, Willmann-Bell, Richmond, Virginia, 1986.

Dunham, J. B, Dunham, D. W. and Warren, W. H., *IOTA Observer's Manual,* (draft copy), 1992.

Espenak, F., *Fifty Year Canon of Solar Eclipses: 1986–2035*, NASA RP-1178, Greenbelt, MD, 1987.

Explanatory Supplement to the Astronomical Ephemeris and the American Ephemeris and Nautical Almanac, Her Majesty's Nautical Almanac Office, London, 1974.

Herald, D., "Correcting Predictions of Solar Eclipse Contact Times for the Effects of Lunar Limb Irregularities," *J. Brit. Ast. Assoc.*, 1983, **93**, 6.

Littmann, M., Willcox, K. and Espenak, F. *Totality, Eclipses of the Sun*, Oxford University Press, New York, 1999.

Meeus, J., *Astronomical Formulae for Calculators,* Willmann-Bell, Inc., Richmond, 1982.

Meeus, J., Grosjean, C., and Vanderleen, W., *Canon of Solar Eclipses*, Pergamon Press, New York, 1966.

Morrison, L. V., "Analysis of lunar occultations in the years 1943–1974...," *Astr. J.*, 1979, **75**, 744.

Morrison, L. V., and Appleby, G. M., "Analysis of lunar occultations - III. Systematic corrections to Watts' limb-profiles for the Moon," *Mon. Not. R. Astron. Soc.*, 1981, **196**, 1013.

Stephenson, F. R., *Historical Eclipses and Earth's Rotation*, Cambridge/New York: Cambridge University Press, 1997 (p.406).

The New International Atlas, Rand McNally, Chicago/New York/San Francisco, 1991.

van den Bergh, G., *Periodicity and Variation of Solar (and Lunar) Eclipses*, Tjeenk Willink, Haarlem, Netherlands, 1955.

Watts, C. B., "The Marginal Zone of the Moon," *Astron. Papers Amer. Ephem.*, 1963, **17**, 1-951.

METEOROLOGY AND TRAVEL

Climates of the World, U. S. Dept. of Commerce, Washington DC, 1972.

Griffiths, J.F., ed.,*World Survey of Climatology, vol 10, Climates of Africa*, Elsevier Pub. Co., New York, 1972.

International Station Meteorological Climate Summary; Vol 4.0 (CDROM), National Climatic Data Center, Asheville, NC, 1996.

Warren, Stephen G., Carole J. Hahn, Julius London, Robert M. Chervin and Roy L. Jenne, *Global Distribution of Total Cloud Cover and Cloud Type Amounts Over Land*, National Center for Atmospheric Research, Boulder, CO., 1986.

World WeatherDisc (CDROM), WeatherDisc Associates Inc., Seattle, WA, 1990.

EYE SAFETY

American Conference of Governmental Industrial Hygienists, "Threshold Limit Values for Chemical Substances and Physical Agents and Biological Exposure Indices," ACGIH, Cincinnati, 1996, p.100.

Chou, B. R., "Safe Solar Filters," *Sky & Telescope*, August 1981, 62:2, 119.

Chou, B. R., "Solar Filter Safety," *Sky & Telescope*, February 1998, 95:2, 119.

Chou, B. R., "Eye safety during solar eclipses - myths and realities," in Z. Madourian & M. Stavinschi (eds.) *Theoretical and Observational Problems Related to Solar Eclipses, Proceedings of a NATO Advanced Research Workshop*. Kluwer Academic Publishers, Dordrecht, 1996, pp.243-247.

Chou, B. R. and Krailo M. D., "Eye injuries in Canada following the total solar eclipse of 26 February 1979," *Can. J. Optometry*, 1981, 43(1):40.

Del Priore, L. V., "Eye damage from a solar eclipse" in M. Littman and K. Willcox, *Totality: Eclipses of the Sun*, University of Hawaii Press, Honolulu, 1991, p. 130.

Marsh, J. C. D., "Observing the Sun in Safety," *J. Brit. Ast. Assoc.*, 1982, **92**, 6.

Penner, R. and McNair, J. N., "Eclipse blindness - Report of an epidemic in the military population of Hawaii," *Am. J. Ophthalmology*, 1966, 61:1452.

Pitts D. G., "Ocular effects of radiant energy," in D. G. Pitts & R. N. Kleinstein (eds.) *Environmental Vision: Interactions of the Eye, Vision and the Environment*, Butterworth-Heinemann, Toronto, 1993, p. 151.

FURTHER READING

Allen, D., and Allen, C., *Eclipse*, Allen & Unwin, Sydney, 1987.

Astrophotography Basics, Kodak Customer Service Pamphlet P150, Eastman Kodak, Rochester, 1988.

Brewer, B., *Eclipse*, Earth View, Seattle, 1991.

Covington, M., *Astrophotography for the Amateur*, Cambridge University Press, Cambridge, 1988.

Espenak, F., "Total Eclipse of the Sun," *Petersen's PhotoGraphic*, June 1991, p. 32.

Fiala, A. D., DeYoung, J. A., and Lukac, M. R., *Solar Eclipses, 1991–2000*, USNO Circular No. 170, U. S. Naval Observatory, Washington, DC, 1986.

Golub, L., and Pasachoff, J. M., *The Solar Corona*, Cambridge University Press, Cambridge, 1997.

Harris, J., and Talcott, R., *Chasing the Shadow,* Kalmbach Pub., Waukesha, 1994.

Littmann, M., Willcox, K. and Espenak, F. *Totality, Eclipses of the Sun*, Oxford University Press, New York, 1999.

Lowenthal, J., *The Hidden Sun: Solar Eclipses and Astrophotography*, Avon, New York, 1984.

Mucke, H., and Meeus, J., *Canon of Solar Eclipses: –2003 to +2526*, Astronomisches Büro, Vienna, 1983.

North, G., *Advanced Amateur Astronomy*, Edinburgh University Press, 1991.

Oppolzer, T. R. von, *Canon of Eclipses*, Dover Publications, New York, 1962.

Ottewell, G., *The Under-Standing of Eclipses*, Astronomical Workshop, Greenville, SC, 1991.

Pasachoff, J. M., "Solar Eclipses and Public Education," International Astronomical Union Colloquium #162: New Trends in Teaching Astronomy, D. McNally, ed., London 1997, in press.

Pasachoff, J. M., and Covington, M., *Cambridge Guide to Eclipse Photography*, Cambridge University Press, Cambridge and New York, 1993.

Pasachoff, J. M., and Menzel, D. H., *Field Guide to the Stars and Planets*, 3rd edition, Houghton Mifflin, Boston, 1992.

Reynolds, M. D. and Sweetsir, R. A., *Observe Eclipses*, Astronomical League, Washington, DC, 1995.

Sherrod, P. C., *A Complete Manual of Amateur Astronomy*, Prentice-Hall, 1981.

Zirker, J. B., *Total Eclipses of the Sun*, Princeton University Press, Princeton, 1995.

FIGURES

Total Solar Eclipse of 2001 Jun 21

FIGURE 1: ORTHOGRAPHIC PROJECTION MAP OF THE ECLIPSE PATH

Geocentric Conjunction = 11:57:49.1 UT   J.D. = 2452081.998485
Greatest Eclipse = 12:03:41.3 UT   J.D. = 2452082.002561

Eclipse Magnitude = 1.04953   Gamma = -0.57014

Saros Series = 127   Member = 57 of 82

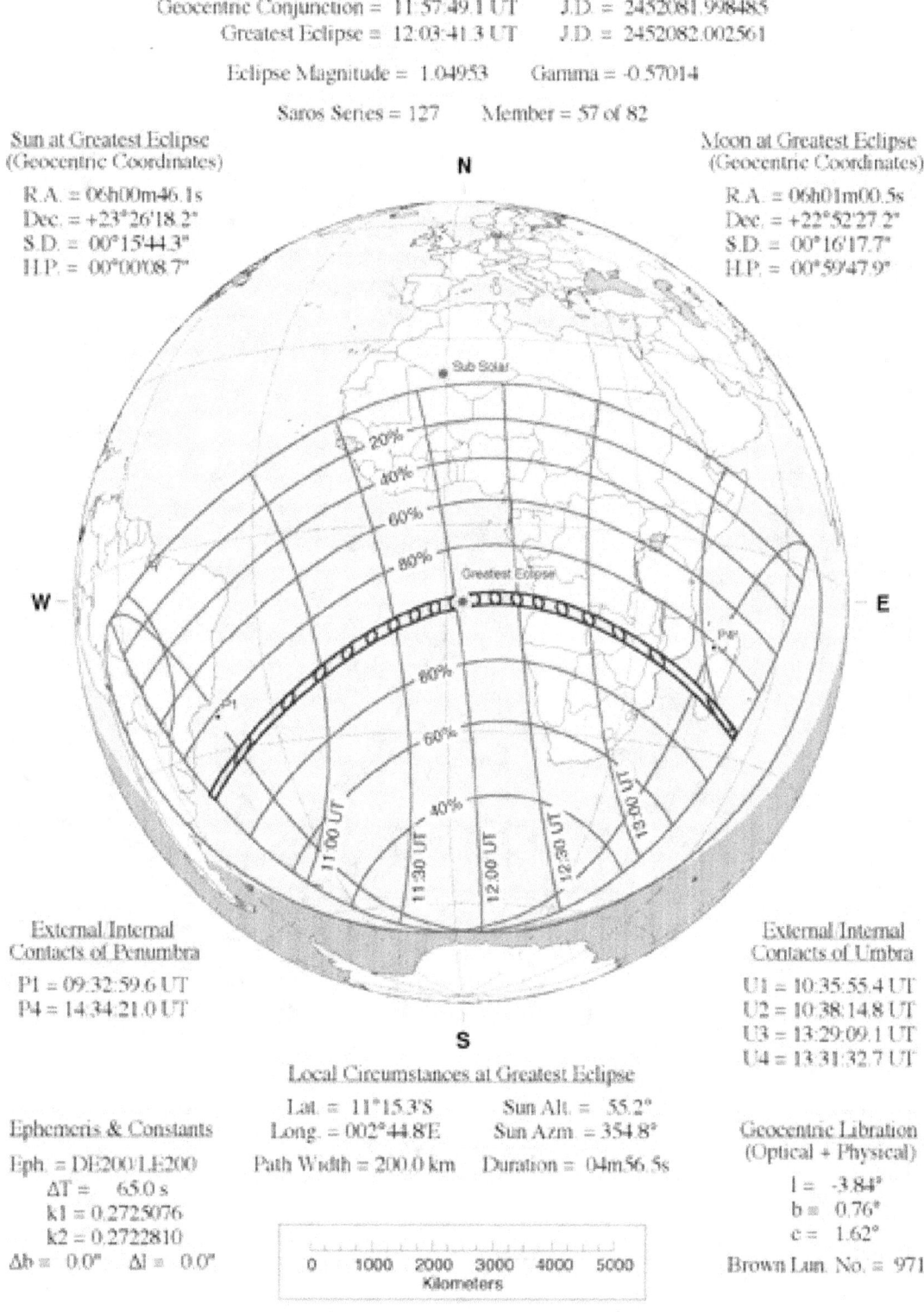

Sun at Greatest Eclipse
(Geocentric Coordinates)

R.A. = 06h00m46.1s
Dec. = +23°26'18.2"
S.D. = 00°15'44.3"
H.P. = 00°00'08.7"

Moon at Greatest Eclipse
(Geocentric Coordinates)

R.A. = 06h01m00.5s
Dec. = +22°52'27.2"
S.D. = 00°16'17.7"
H.P. = 00°59'47.9"

External/Internal
Contacts of Penumbra

P1 = 09:32:59.6 UT
P4 = 14:34:21.0 UT

External/Internal
Contacts of Umbra

U1 = 10:35:55.4 UT
U2 = 10:38:14.8 UT
U3 = 13:29:09.1 UT
U4 = 13:31:32.7 UT

Local Circumstances at Greatest Eclipse

Lat. = 11°15.3'S   Sun Alt. = 55.2°
Long. = 002°44.8'E   Sun Azm. = 354.8°
Path Width = 200.0 km   Duration = 04m56.5s

Ephemeris & Constants

Eph. = DE200/LE200
ΔT = 65.0 s
k1 = 0.2725076
k2 = 0.2722810
Δb = 0.0"   Δl = 0.0"

Geocentric Libration
(Optical + Physical)

l = -3.84°
b = 0.76°
c = 1.62°

Brown Lun. No. = 971

0 1000 2000 3000 4000 5000
Kilometers

F. Espenak, NASA/GSFC - 1999 Sep 7

Total Solar Eclipse of 2001 Jun 21

FIGURE 2: STEREOGRAPHIC PROJECTION MAP OF THE ECLIPSE PATH

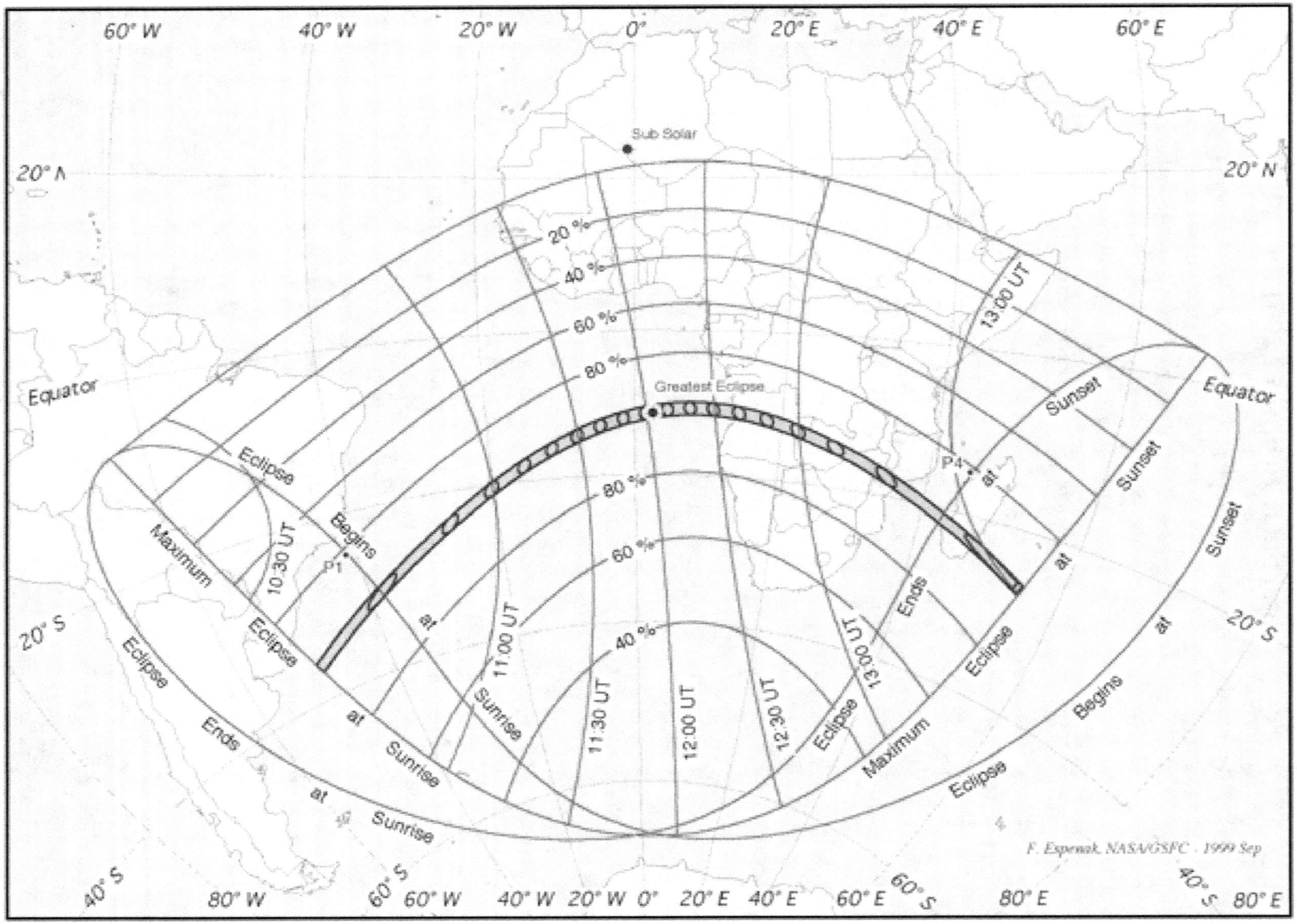

Total Solar Eclipse of 2001 June 21

FIGURE 3: THE ECLIPSE PATH THROUGH AFRICA

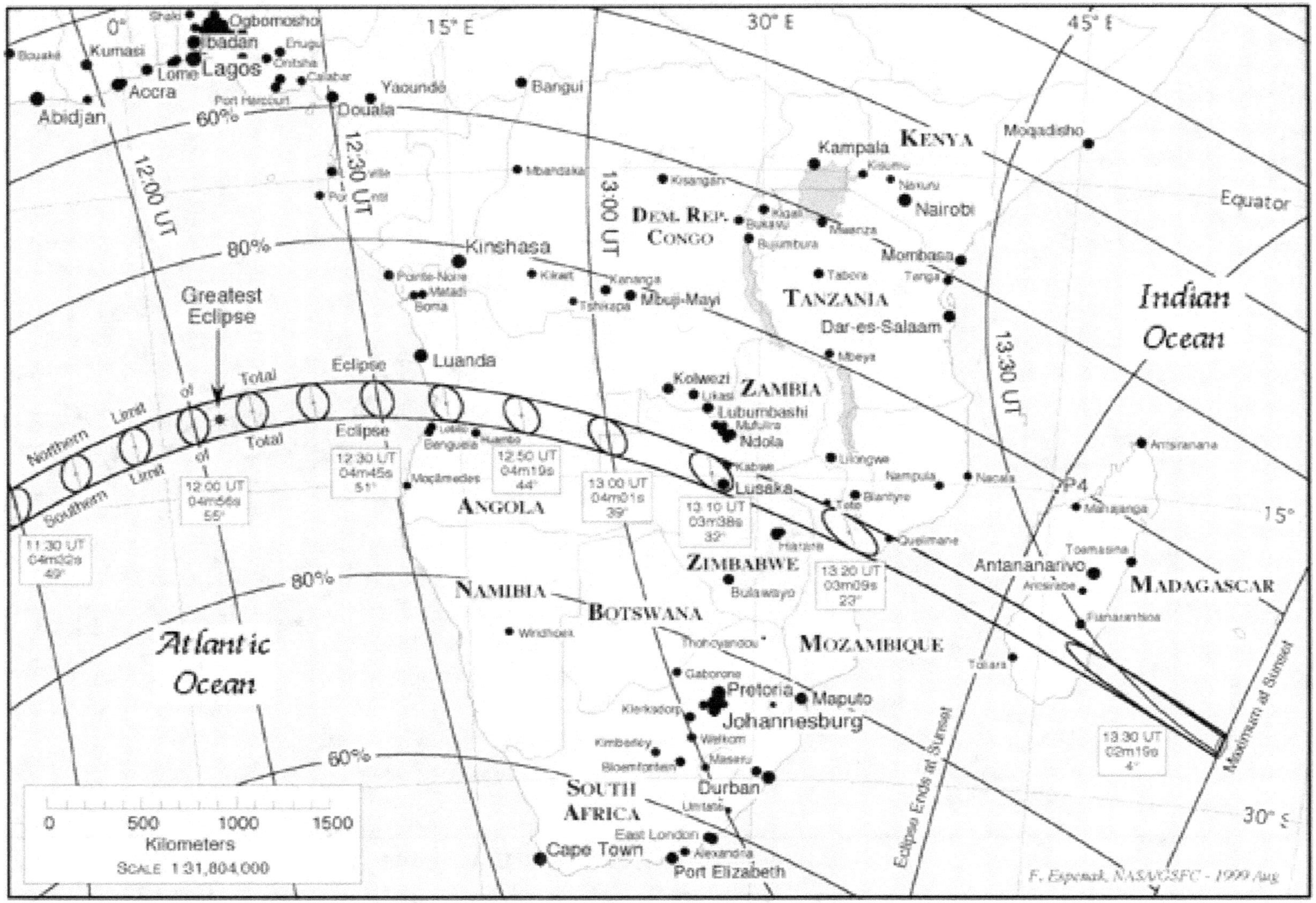

Total Solar Eclipse of 2001 June 21

FIGURE 4: THE ECLIPSE PATH THROUGH ANGOLA

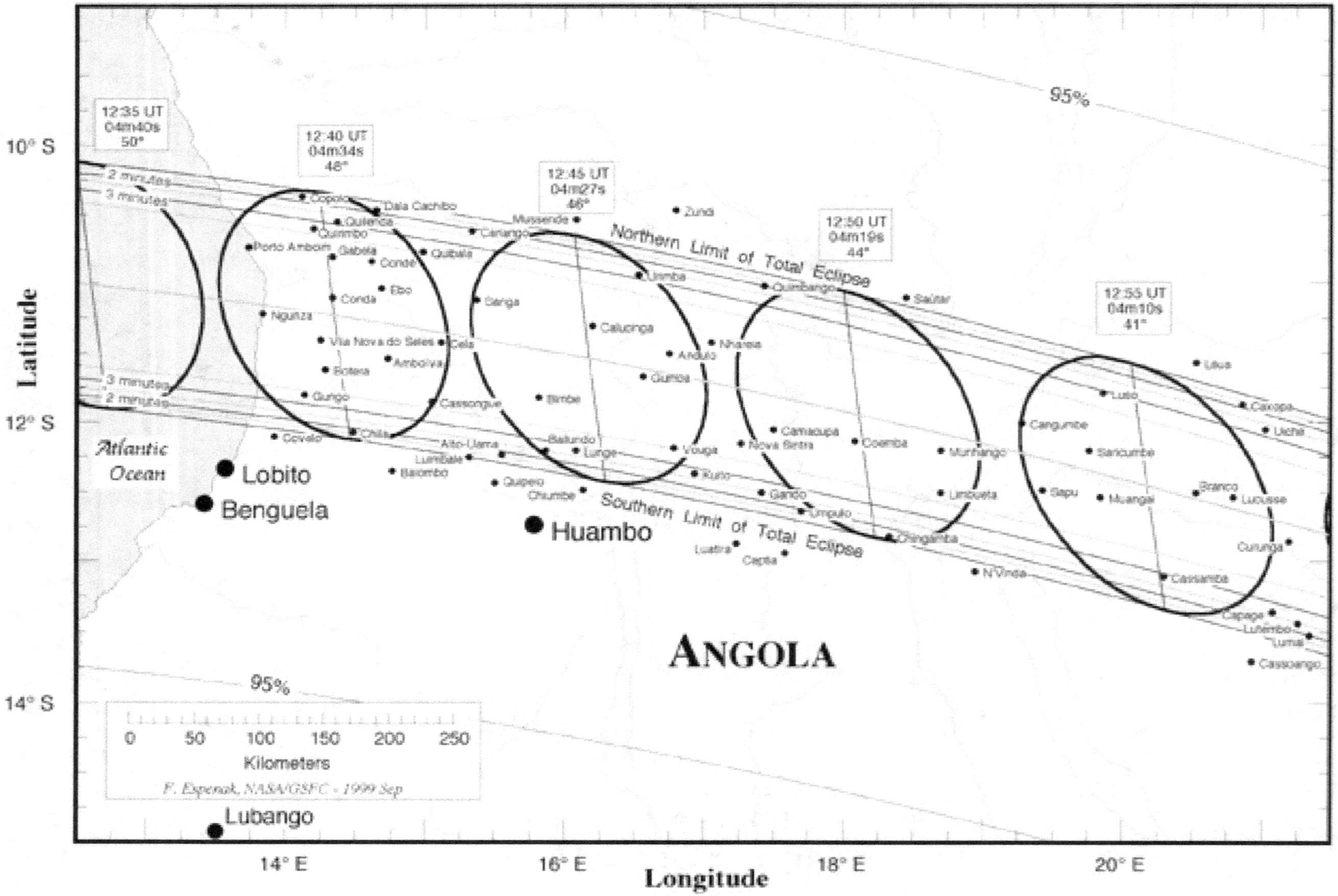

Total Solar Eclipse of 2001 June 21

Figure 5: The Eclipse Path Through Zambia

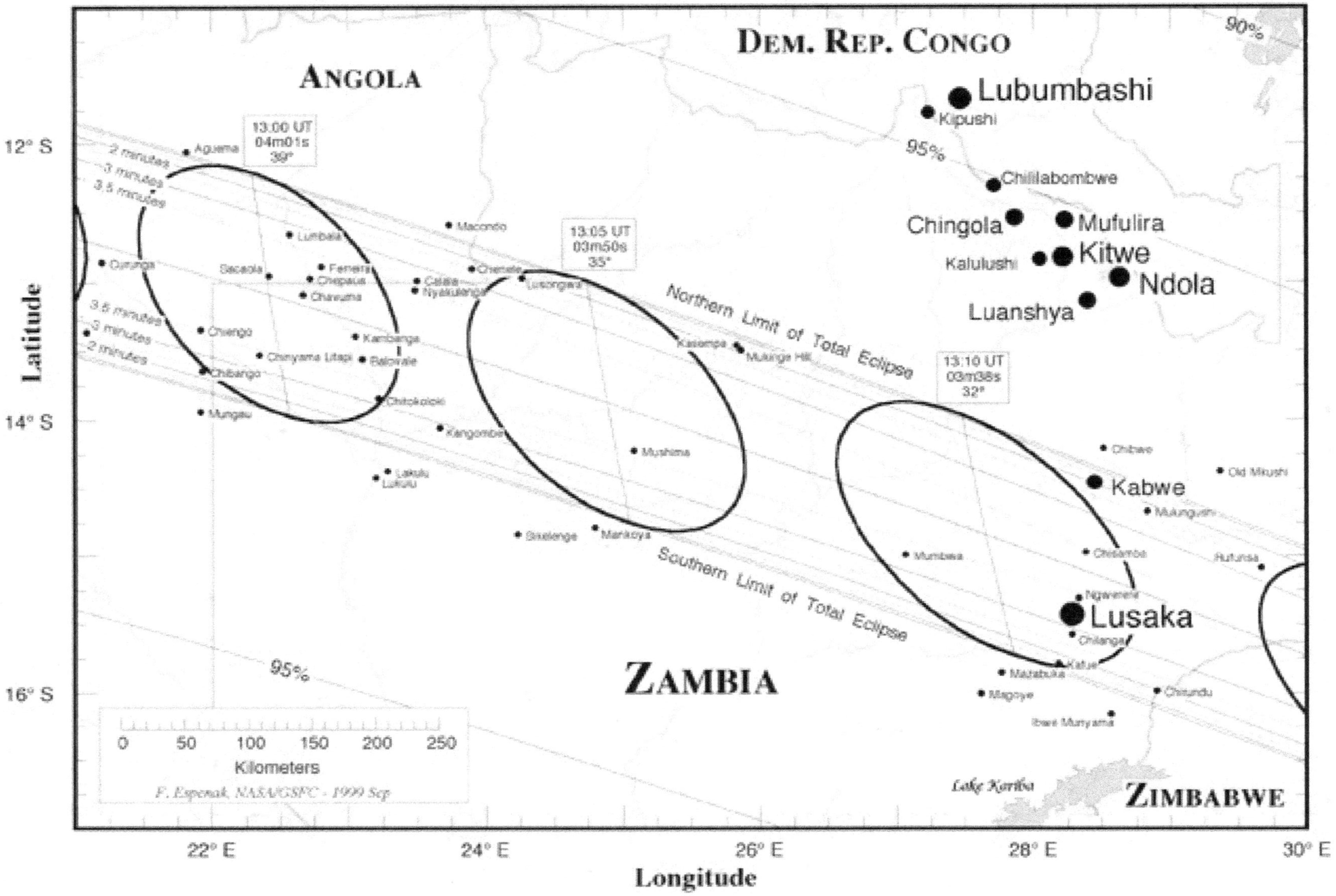

37

Total Solar Eclipse of 2001 June 21

FIGURE 6: THE ECLIPSE PATH THROUGH ZIMBABWE AND MOZAMBIQUE

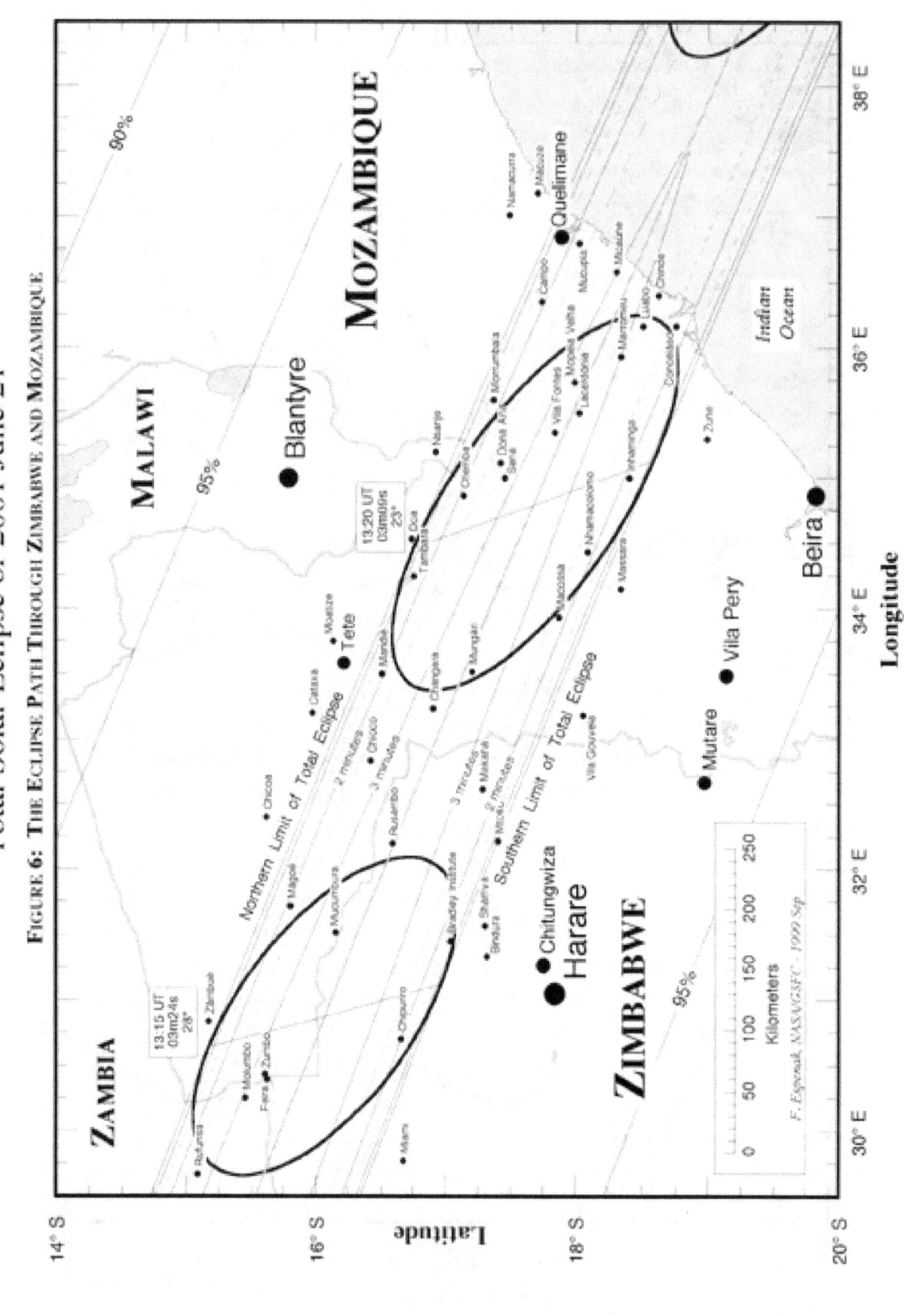

Total Solar Eclipse of 2001 June 21

FIGURE 7: THE ECLIPSE PATH THROUGH MADAGASCAR

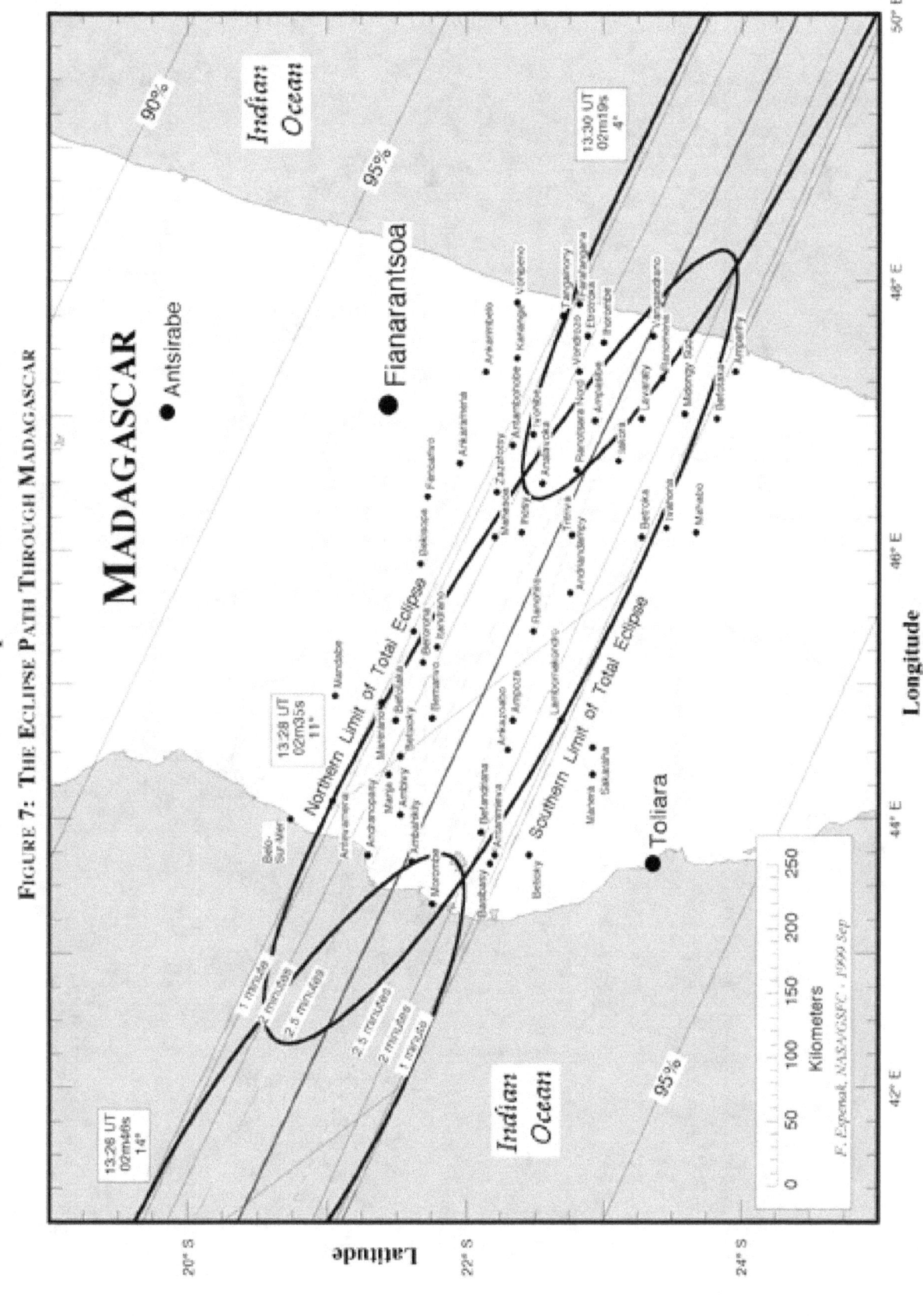

Total Solar Eclipse of 2001 Jun 21

FIGURE 8: THE LUNAR LIMB PROFILE AT 13:00 UT

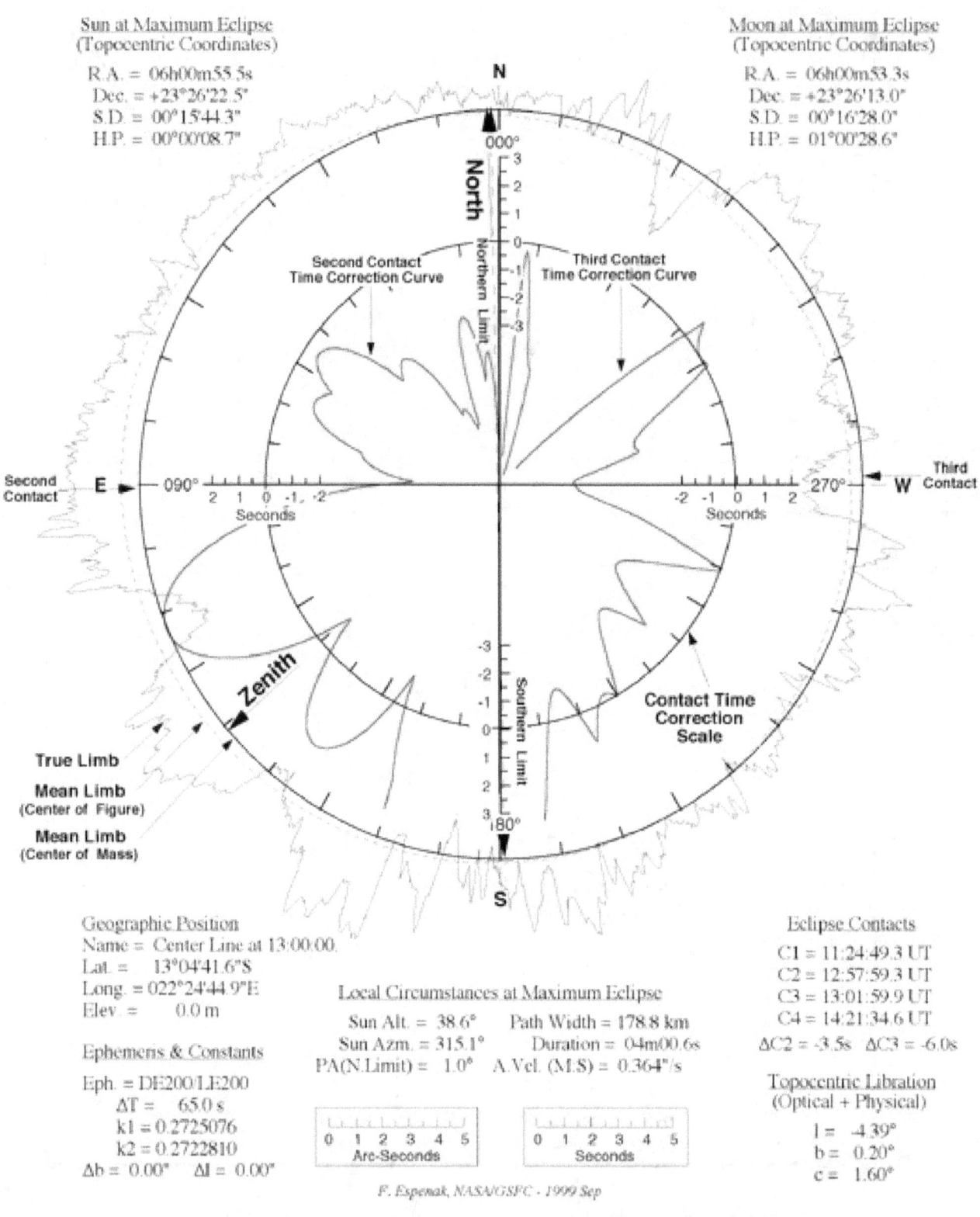

Sun at Maximum Eclipse
(Topocentric Coordinates)

R.A. = 06h00m55.5s
Dec. = +23°26'22.5"
S.D. = 00°15'44.3"
H.P. = 00°00'08.7"

Moon at Maximum Eclipse
(Topocentric Coordinates)

R.A. = 06h00m53.3s
Dec. = +23°26'13.0"
S.D. = 00°16'28.0"
H.P. = 01°00'28.6"

Second Contact
Time Correction Curve

Third Contact
Time Correction Curve

Second Contact

Third Contact

Contact Time Correction Scale

True Limb

Mean Limb
(Center of Figure)

Mean Limb
(Center of Mass)

Geographic Position
Name = Center Line at 13:00:00.
Lat. = 13°04'41.6"S
Long. = 022°24'44.9"E
Elev = 0.0 m

Ephemeris & Constants
Eph. = DE200/LE200
ΔT = 65.0 s
k1 = 0.2725076
k2 = 0.2722810
Δb = 0.00"   Δl = 0.00"

Local Circumstances at Maximum Eclipse

Sun Alt. = 38.6°   Path Width = 178.8 km
Sun Azm. = 315.1°   Duration = 04m00.6s
PA(N.Limit) = 1.0°   A.Vel. (M:S) = 0.364"/s

Arc-Seconds

Seconds

F. Espenak, NASA/GSFC - 1999 Sep

Eclipse Contacts

C1 = 11:24:49.3 UT
C2 = 12:57:59.3 UT
C3 = 13:01:59.9 UT
C4 = 14:21:34.6 UT

ΔC2 = -3.5s   ΔC3 = -6.0s

Topocentric Libration
(Optical + Physical)

l = -4.39°
b = 0.20°
c = 1.60°

cor.C2 = 12:57:55.8 UT (-3.5s)   cor.C3 = 13:01:53.9 UT (-6.0s)

Total Solar Eclipse of 2001 June 21

FIGURE 9: LIMB PROFILE EFFECTS ON THE DURATION OF TOTALITY

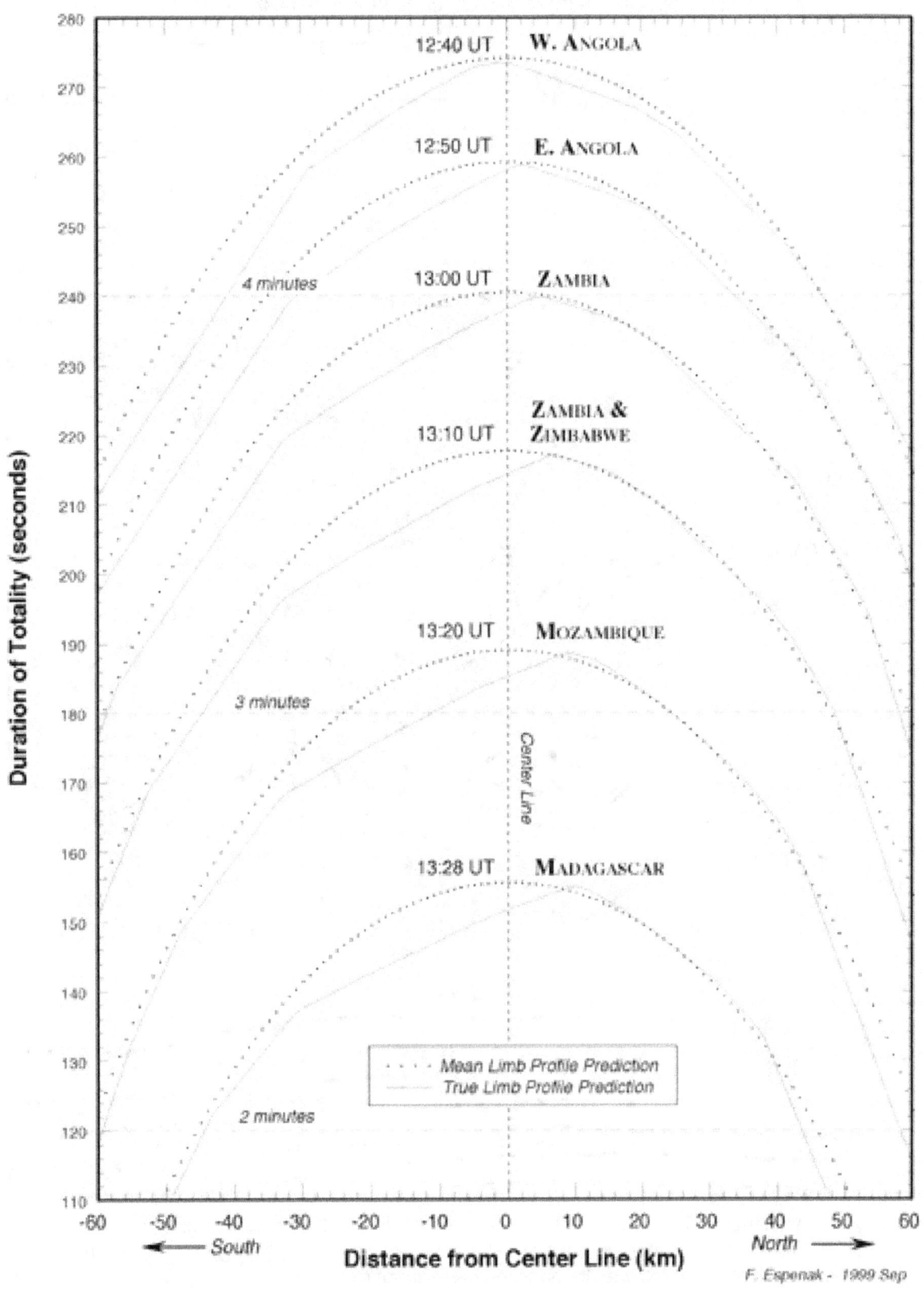

F. Espenak - 1999 Sep

Total Solar Eclipse of 2001 June 21

Figure 10: Mean Pressure in June for Africa

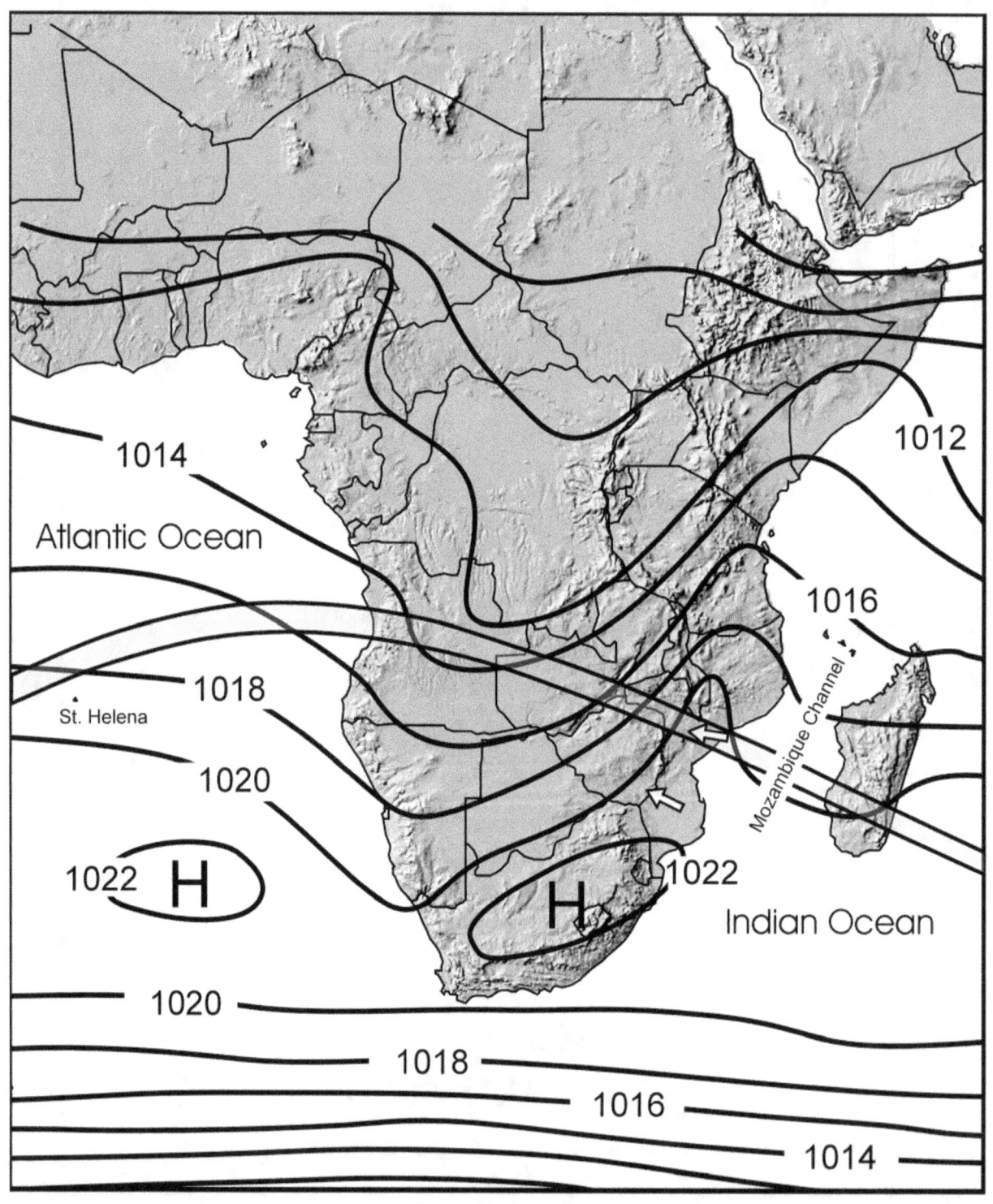

Figure 10 shows the climatological and topographic setting for the 2001 eclipse across Africa. The contours depict the mean June pressure pattern in millibars. The semi-permanent high pressure anticyclones over South Africa and adjacent waters contribute to the suppression of cloudiness over the southern continent. The small arrows over Mozambique show the main route for cloud to move into the mid continent, along the courses of the Limpopo and Zambezi Rivers.

Total Solar Eclipse of 2001 June 21

Figure 11: Mean Cloud Cover in June for Southern Africa

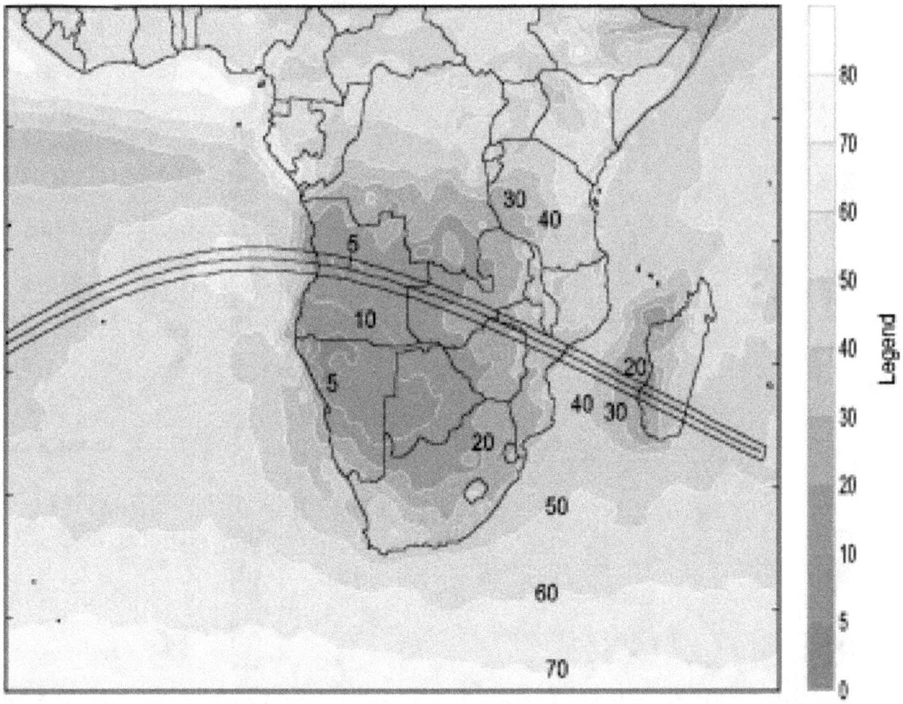

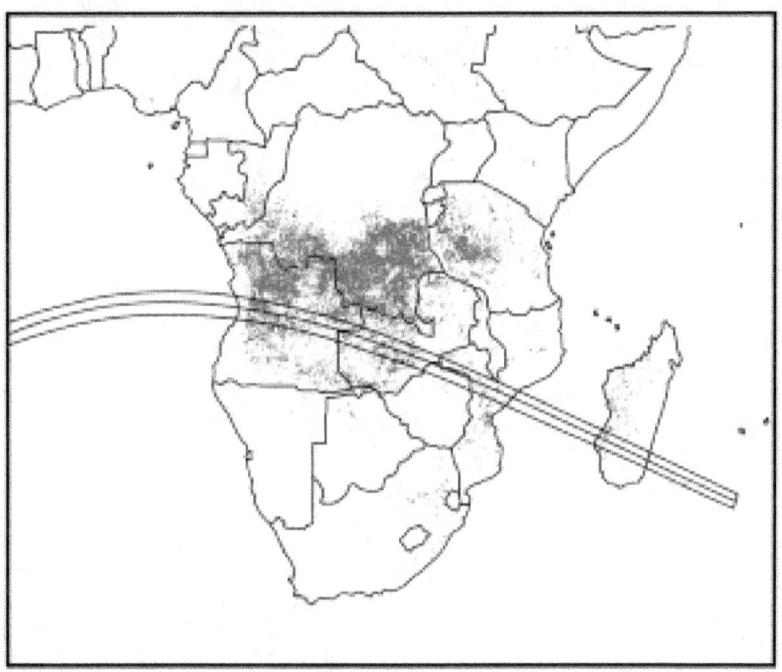

Mean June daytime cloud cover (%) over southern Africa. The data for this chart were collected from satellite observations on a 1 degree by 1 degree grid (100 km x 100 km) over a 13 year period, from 1983 to 1994. Darker areas represent lower cloud amounts and thus more favorable viewing conditions for the 2001 eclipse.

Figure 12: Biomass Burning in June for Southern Africa

The extent of biomass burning in June 1993 along the path of the eclipse is shown above. The data were collected from satellite observations as part of the Global Fire Monitoring Program sponsored, in part, by NASA and NOAA. While these data come from only one year of observation, the pattern is typical of the seasonal burning over southern Africa. The result of these fires can be seen in the statistics for the frequency of smoke and haze for Lusaka, Zambia and other sites along the path.

Total Solar Eclipse of 2001 June 21

Figure 13: Cloud Cover Statistics Along the Eclipse Path

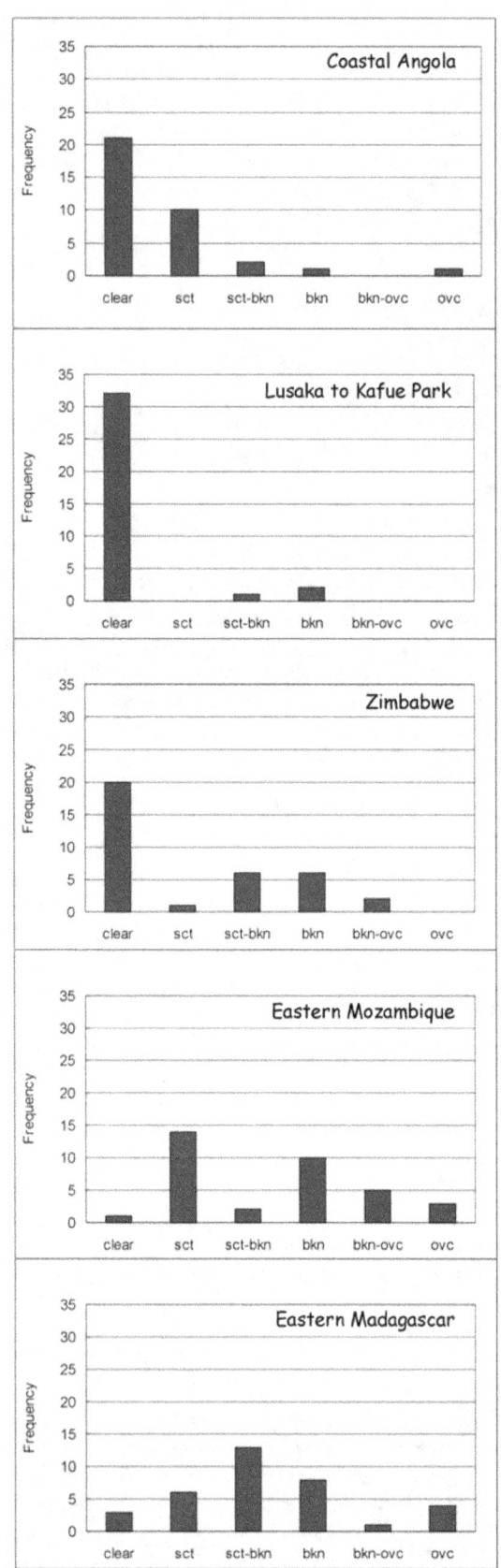

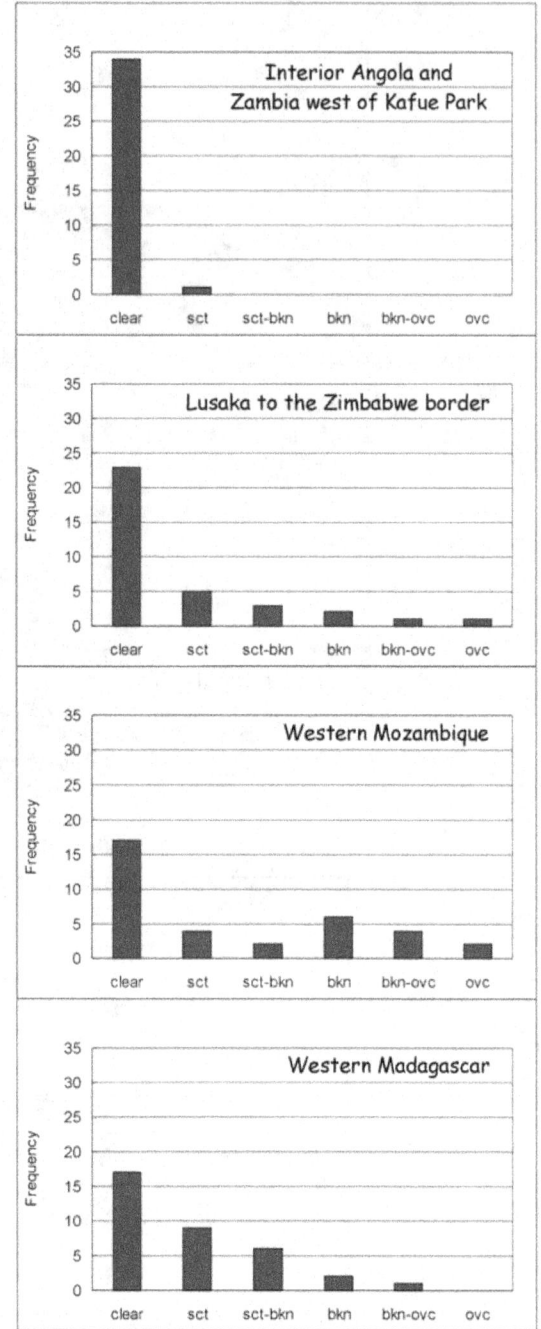

Cloud cover statistics for selected sites along the eclipse track. These statistics were derived from geostationary satellite images acquired at 1800 UTC between June 1 and July 8, 1998. A total of 35 days were analyzed. A sample such as this, taken from only one year of imagery, is not statistically significant, but may represent the relative trend in cloudiness between sites.

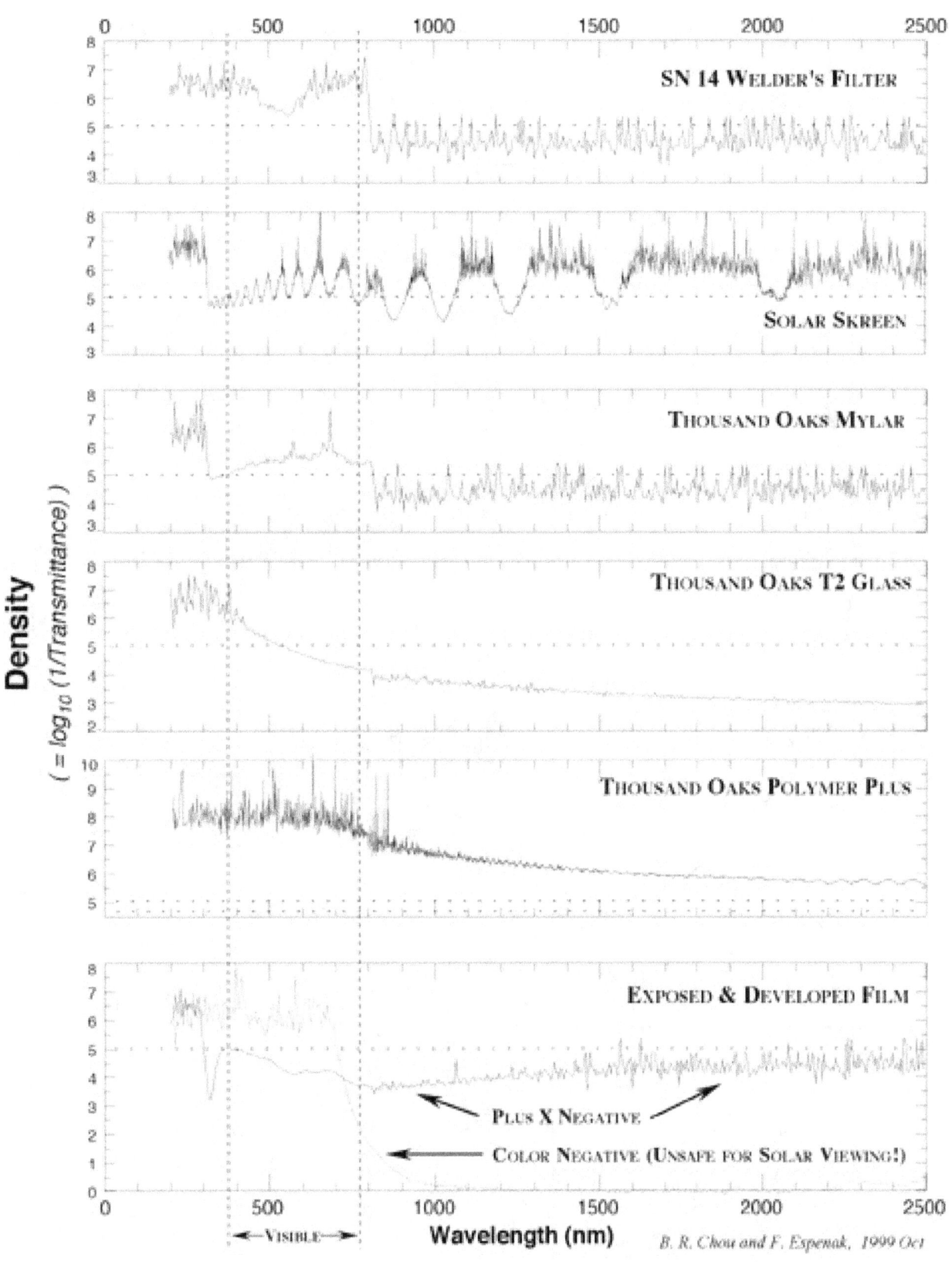

Total Solar Eclipse of 2001 June 21

FIGURE 15: THE SKY DURING TOTALITY AS SEEN FROM CENTER LINE AT 13:00 UT

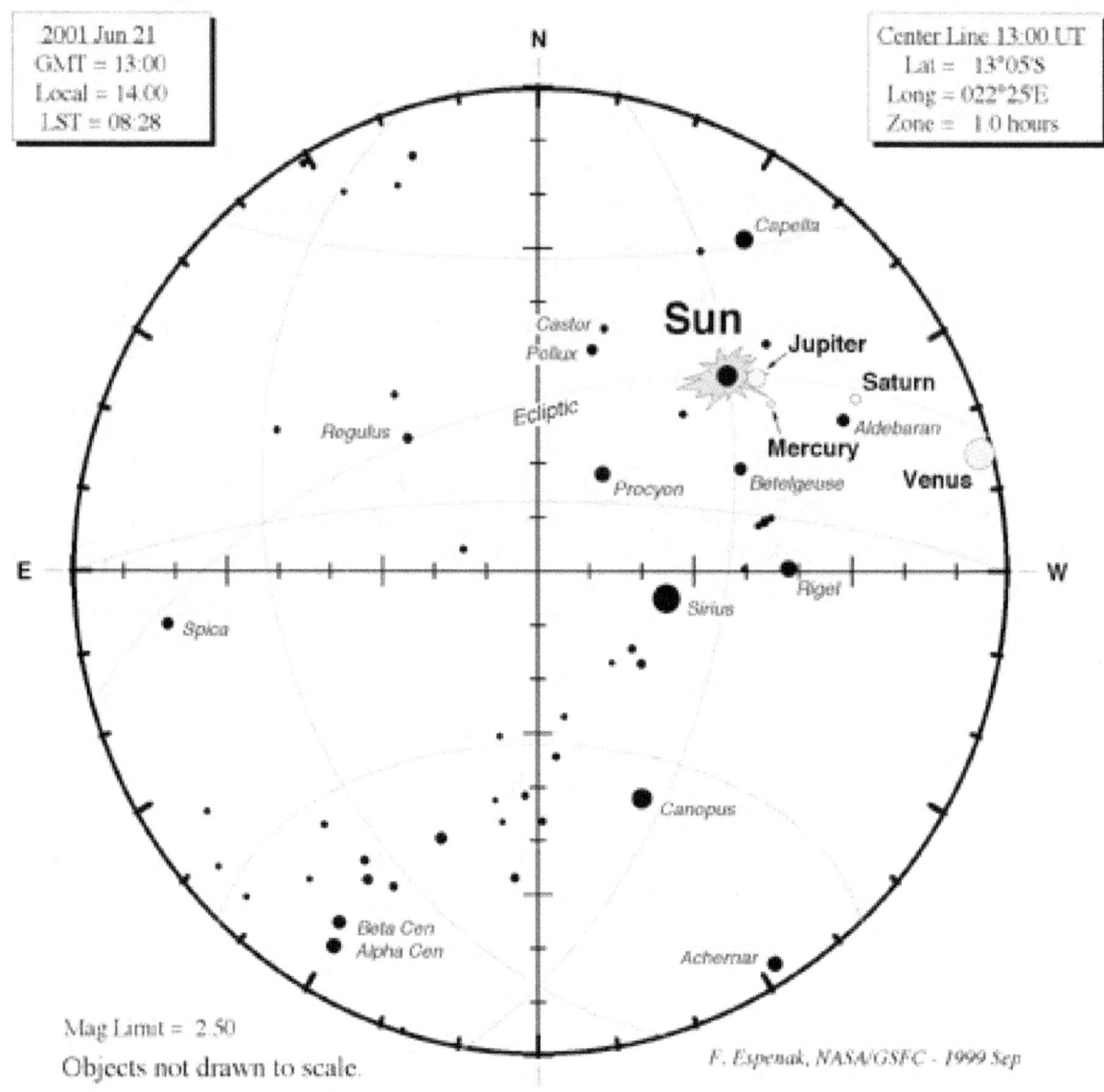

2001 Jun 21
GMT = 13:00
Local = 14:00
LST = 08:28

Center Line 13:00 UT
Lat = 13°05'S
Long = 022°25'E
Zone = 1.0 hours

Mag Limit = 2.50

Objects not drawn to scale.

F. Espenak, NASA/GSFC - 1999 Sep

Figure 15: The sky during totality as seen from the center line in Zambia at 13:00 UT. The most conspicuous planet visible will be Jupiter (m=−1.5) just 5° west of the Sun. Although Mercury (m=+2.7) is 8.6° west of the Sun, it is quite faint and will be difficult to spot. Saturn (m=+0.3) should prove an easier target 22.6° west of the Sun. Venus (m=−3.3) is brightest of all, but sets before totality for locations east of central Zambia. The northwestern sky will be dominated by the bright stars of winter, including Capella (m=+0.08), Aldebaran (m=+0.85v), Procyon (m=+0.38), Betelgeuse (m=+0.5v), Rigel (m=+0.12), and Sirius (m=−1.46). Other bright stars which may also be visible include Spica (m=+1.0v), Regulus (m=+1.35), and Canopus (m=−0.72). Star visibility requires a very dark and cloud free sky during totality.

For sky maps from other locations along the path of totality, see the special 2001 eclipse web site:
http://sunearth.gsfc.nasa.gov/eclipse/TSE2001T/SE2001.html

FIGURE 16: ANNULAR ECLIPSE PATHS THROUGH MEXICO & CENTRAL AMERICA

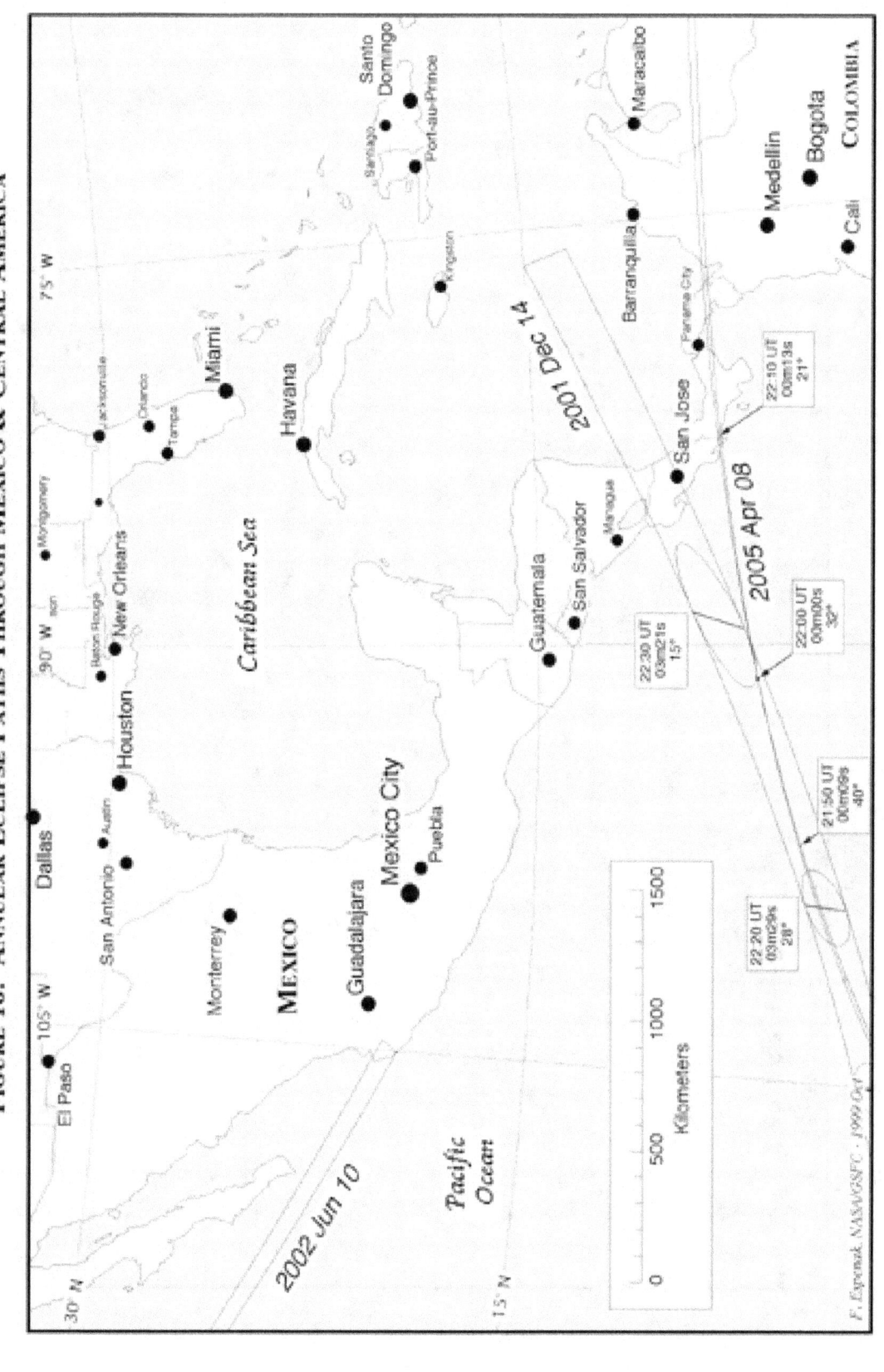

47

Total Solar Eclipse of 2002 December 04

FIGURE 17: THE ECLIPSE PATH THROUGH AFRICA

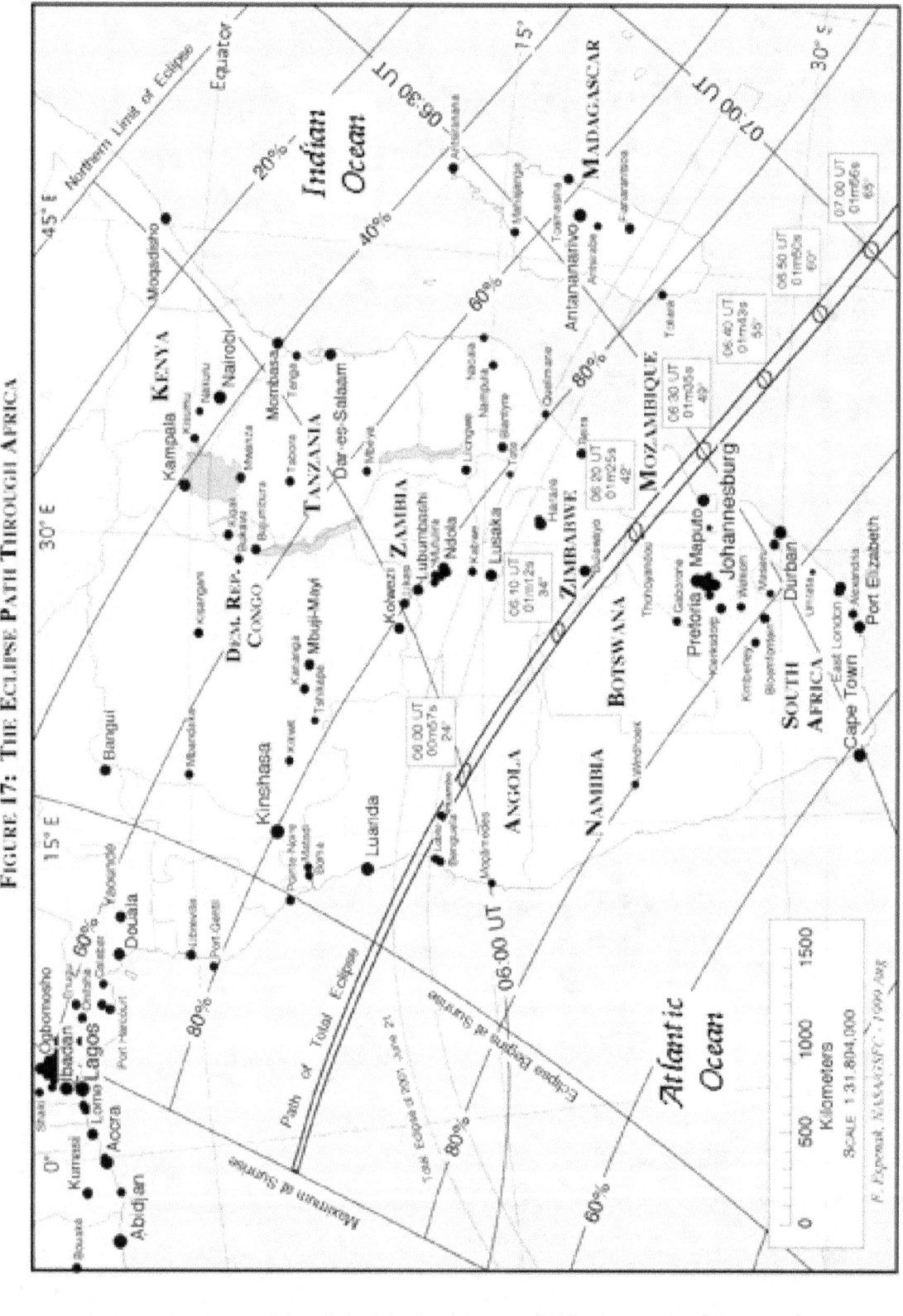

Total Solar Eclipse of 2002 December 04

FIGURE 18: THE ECLIPSE PATH THROUGH AUSTRALIA

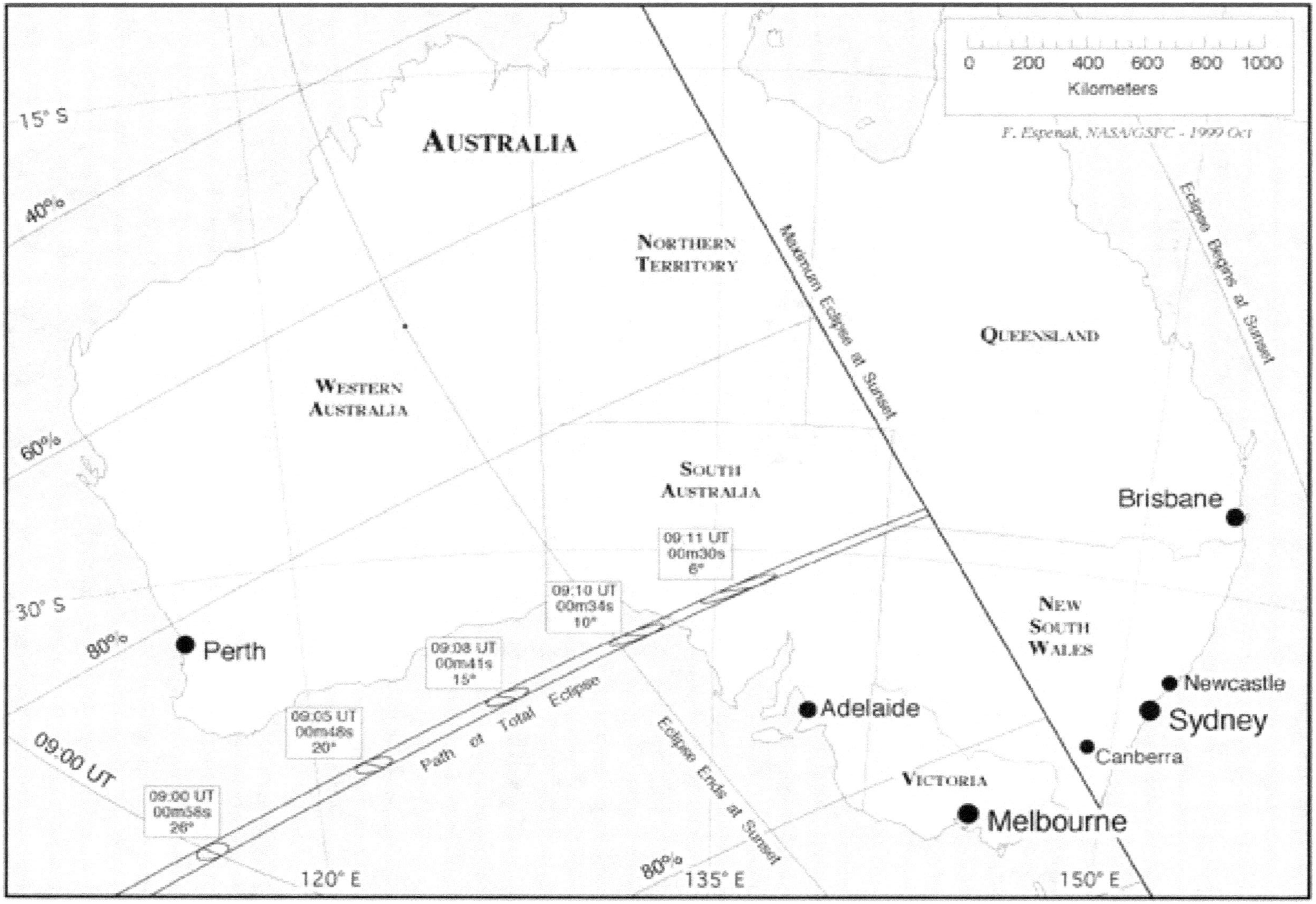

TOTAL SOLAR ECLIPSE OF 2001 JUNE 21

TABLES

TABLE 1

ELEMENTS OF THE TOTAL SOLAR ECLIPSE OF 2001 JUNE 21

<u>Geocentric Conjunction</u>   11:58:54.06 TDT   J.D. = 2452081.999237
<u>of Sun & Moon in R.A.:</u>   (=11:57:49.06 UT)

<u>Instant of</u>   12:04:46.26 TDT   J.D. = 2452082.003313
<u>Greatest Eclipse:</u>   (=12:03:41.26 UT)

<u>Geocentric Coordinates of Sun & Moon at Greatest Eclipse (DE200/LE200):</u>

Sun: R.A. = 06h00m46.131s   Moon: R.A. = 06h01m00.453s
           Dec. =+23°26′18.20″   Dec. =+22°52′27.17″
  Semi-Diameter =   15′44.26″   Semi-Diameter =   16′17.69″
    Eq.Hor.Par. =   08.65″   Eq.Hor.Par. = 0°59′47.93″
         Δ R.A. =   10.406s/h   Δ R.A. =   156.807s/h
         Δ Dec. =   -0.18″/h   Δ Dec. =   196.15″/h

<u>Lunar Radius</u>   k1 = 0.2725076 (Penumbra)   <u>Shift in</u>   Δb = 0.00″
 <u>Constants:</u>   k2 = 0.2722810 (Umbra)   <u>Lunar Position:</u>   Δl = 0.00″

<u>Geocentric Libration:</u>   l = -3.8°   Brown Lun. No. = 971
(Optical + Physical)   b =   0.7°   Saros Series = 127 (57/82)
                       c =   2.0°   Ephemeris = (DE200/LE200)

<u>Eclipse Magnitude</u> = 1.04953   <u>Gamma</u> =-0.57014   ΔT =   65.0 s

<u>Polynomial Besselian Elements for:</u>   2001 Jun 21   12:00:00.0 TDT   (=t0)

n	x	y	d	l_1	l_2	μ
0	0.0103551	-0.5718306	23.4397678	0.5369922	-0.0091193	359.564514
1	0.5653870	0.0551258	-0.0001831	-0.0000944	-0.0000939	14.999194
2	0.0000292	-0.0001339	-0.0000056	-0.0000121	-0.0000121	0.000000
3	-0.0000089	-0.0000010	0.0000000	0.0000000	0.0000000	0.000000

Tan f_1 = 0.0046005   Tan f_2 = 0.0045776

At time 't1' (decimal hours), each Besselian element is evaluated by:

$$a = a_0 + a_1{*}t + a_2{*}t^2 + a_3{*}t^3 \qquad (\text{or } a = \Sigma\ [a_n{*}t^n];\ n = 0\ \text{to}\ 3)$$

where: a = x, y, d, l_1, l_2, or μ
       t = t_1 - t_0 (decimal hours) and t_0 = 12.000 TDT

The Besselian elements were derived from a least-squares fit to elements calculated at five uniformly spaced times over a six hour period centered at t_0. Thus
the elements are valid over the period 9.00 ≤ t_1 ≤ 15.00 TDT.

   Saros Series 127: Member 57 of 82 eclipses in series.

TABLE 2

SHADOW CONTACTS AND CIRCUMSTANCES
TOTAL SOLAR ECLIPSE OF 2001 JUNE 21

$$\Delta T = \quad 65.0 \text{ s}$$
$$= 000°16'17.7"E$$

		Terrestrial Dynamical Time h m s	Latitude	Ephemeris Longitude†	True Longitude*
External/Internal Contacts of Penumbra:	P1	09:34:04.6	25°04.6'S	041°22.8'W	041°06.5'W
	P4	14:35:26.0	14°57.4'S	044°55.7'E	045°12.0'E
Extreme North/South Limits of Penumbral Path:	N1	10:29:28.2	07°21.7'S	063°43.3'W	063°27.0'W
	S1	13:40:03.3	02°53.3'N	066°40.6'E	066°56.9'E
External/Internal Contacts of Umbra:	U1	10:37:00.4	36°18.6'S	050°14.3'W	049°58.0'W
	U2	10:39:19.8	36°56.2'S	050°22.5'W	050°06.2'W
	U3	13:30:14.1	27°05.1'S	055°04.0'E	055°20.3'E
	U4	13:32:37.7	26°25.2'S	054°50.4'E	055°06.7'E
Extreme North/South Limits of Umbral Path:	N1	10:37:40.6	36°04.8'S	050°34.0'W	050°17.7'W
	S1	10:38:40.5	37°09.7'S	050°02.8'W	049°46.6'W
	N2	13:31:56.2	26°10.6'S	055°08.9'E	055°25.1'E
	S2	13:30:54.7	27°19.6'S	054°45.8'E	055°02.1'E
Extreme Limits of Center Line:	C1	10:38:09.9	36°37.2'S	050°18.5'W	050°02.3'W
	C2	13:31:26.1	26°45.0'S	054°57.3'E	055°13.6'E
Instant of Greatest Eclipse:	G0	12:04:46.3	11°15.3'S	002°28.5'E	002°44.8'E

Circumstances at
Greatest Eclipse: Sun's Altitude = 55.2°   Path Width = 200.0 km
                  Sun's Azimuth = 354.8°   Central Duration = 04m56.5s

---

† Ephemeris Longitude is the terrestrial dynamical longitude assuming a
uniformly rotating Earth.
* True Longitude is calculated by correcting the Ephemeris Longitude for
the non-uniform rotation of Earth.
(T.L. = E.L. + 1.002738*ΔT/240, where ΔT(in seconds) = TDT - UT)

Note: Longitude is measured positive to the East.

Since ΔT is not known in advance, the value used in the predictions is an
extrapolation based on pre-2001 measurements. Nevertheless, the actual
value is expected to fall within ±0.5 seconds of the estimated ΔT used here.

TABLE 3

PATH OF THE UMBRAL SHADOW
TOTAL SOLAR ECLIPSE OF 2001 JUNE 21

Universal Time	Northern Limit Latitude	Northern Limit Longitude	Southern Limit Latitude	Southern Limit Longitude	Center Line Latitude	Center Line Longitude	Sun Alt °	Path Width km	Central Durat.
Limits	36°04.8'S	050°17.7'W	37°09.7'S	049°46.6'W	36°37.2'S	050°02.3'W	0	127	02m06.1s
10:40	29°34.2'S	037°18.4'W	31°47.6'S	038°43.6'W	30°38.9'S	037°56.6'W	12	140	02m34.3s
10:45	26°01.1'S	030°51.3'W	27°52.7'S	031°22.0'W	26°56.2'S	031°05.1'W	20	148	02m55.4s
10:50	23°33.0'S	026°30.9'W	25°18.7'S	026°43.5'W	24°25.3'S	026°36.2'W	26	155	03m11.6s
10:55	21°35.4'S	023°06.0'W	23°18.7'S	023°09.0'W	22°26.6'S	023°06.8'W	30	160	03m25.4s
11:00	19°57.0'S	020°13.6'W	21°39.2'S	020°10.6'W	20°47.7'S	020°11.5'W	34	165	03m37.8s
11:05	18°32.2'S	017°42.6'W	20°14.1'S	017°35.5'W	19°22.9'S	017°38.6'W	37	169	03m49.1s
11:10	17°18.0'S	015°26.9'W	19°00.0'S	015°16.9'W	18°08.7'S	015°21.5'W	40	173	03m59.4s
11:15	16°12.4'S	013°22.6'W	17°54.7'S	013°10.4'W	17°03.3'S	013°16.1'W	43	177	04m08.8s
11:20	15°14.0'S	011°26.8'W	16°56.8'S	011°13.1'W	16°05.2'S	011°19.6'W	45	180	04m17.5s
11:25	14°22.0'S	009°37.8'W	16°05.3'S	009°23.0'W	15°13.4'S	009°30.1'W	47	183	04m25.2s
11:30	13°35.5'S	007°54.0'W	15°19.5'S	007°38.5'W	14°27.3'S	007°46.0'W	49	187	04m32.2s
11:35	12°54.2'S	006°14.3'W	14°38.8'S	005°58.3'W	13°46.3'S	006°06.1'W	51	189	04m38.4s
11:40	12°17.6'S	004°37.8'W	14°02.8'S	004°21.5'W	13°10.0'S	004°29.5'W	52	192	04m43.7s
11:45	11°45.5'S	003°03.7'W	13°31.2'S	002°47.4'W	12°38.2'S	002°55.4'W	53	194	04m48.1s
11:50	11°17.5'S	001°31.3'W	13°03.8'S	001°15.1'W	12°10.5'S	001°23.1'W	54	196	04m51.6s
11:55	10°53.7'S	000°00.2'W	12°40.5'S	000°15.9'E	11°46.9'S	000°07.9'E	55	198	04m54.2s
12:00	10°33.8'S	001°30.4'E	12°21.0'S	001°46.1'E	11°27.3'S	001°38.3'E	55	199	04m55.9s
12:05	10°17.8'S	003°00.8'E	12°05.4'S	003°16.2'E	11°11.5'S	003°08.5'E	55	200	04m56.6s
12:10	10°05.8'S	004°31.6'E	11°53.6'S	004°46.6'E	10°59.6'S	004°39.1'E	55	200	04m56.3s
12:15	09°57.6'S	006°03.3'E	11°45.6'S	006°17.8'E	10°51.5'S	006°10.5'E	54	200	04m55.1s
12:20	09°53.4'S	007°36.3'E	11°41.6'S	007°50.4'E	10°47.4'S	007°43.3'E	54	200	04m52.8s
12:25	09°53.4'S	009°11.2'E	11°41.5'S	009°25.0'E	10°47.3'S	009°18.0'E	53	199	04m49.6s
12:30	09°57.6'S	010°48.7'E	11°45.6'S	011°02.2'E	10°51.5'S	010°55.3'E	51	197	04m45.5s
12:35	10°06.2'S	012°29.3'E	11°54.1'S	012°42.7'E	11°00.0'S	012°35.8'E	50	195	04m40.3s
12:40	10°19.7'S	014°14.0'E	12°07.3'S	014°27.3'E	11°13.4'S	014°20.4'E	48	193	04m34.2s
12:45	10°38.3'S	016°03.6'E	12°25.7'S	016°17.2'E	11°31.8'S	016°10.2'E	46	190	04m27.2s
12:50	11°02.7'S	017°59.5'E	12°49.7'S	018°13.7'E	11°56.0'S	018°06.3'E	44	186	04m19.3s
12:55	11°33.4'S	020°03.2'E	13°20.2'S	020°18.3'E	12°26.6'S	020°10.4'E	41	183	04m10.4s
13:00	12°11.6'S	022°16.9'E	13°58.1'S	022°33.6'E	13°04.7'S	022°24.7'E	39	179	04m00.6s
13:05	12°58.7'S	024°43.6'E	14°45.2'S	025°02.7'E	13°51.7'S	024°52.6'E	35	174	03m49.7s
13:10	13°57.0'S	027°28.0'E	15°43.7'S	027°50.9'E	14°50.1'S	027°38.8'E	32	170	03m37.8s
13:15	15°10.0'S	030°37.8'E	16°57.7'S	031°06.9'E	16°03.5'S	030°51.5'E	28	164	03m24.4s
13:20	16°44.8'S	034°27.7'E	18°35.2'S	035°08.0'E	17°39.6'S	034°46.7'E	23	158	03m09.1s
13:25	18°58.2'S	039°33.5'E	20°57.1'S	040°39.7'E	19°56.8'S	040°04.4'E	16	150	02m50.6s
13:30	23°14.4'S	048°58.2'E	00°00.0'N	000°00.0'E	24°53.5'S	051°05.3'E	4	136	02m19.3s
Limits	26°10.6'S	055°25.1'E	27°19.6'S	055°02.1'E	26°45.0'S	055°13.6'E	0	131	02m09.8s

TABLE 4

PHYSICAL EPHEMERIS OF THE UMBRAL SHADOW
TOTAL SOLAR ECLIPSE OF 2001 JUNE 21

Universal Time	Center Line Latitude	Center Line Longitude	Diameter Ratio	Eclipse Obscur.	Sun Alt °	Sun Azm °	Path Width km	Major Axis km	Minor Axis km	Umbra Veloc. km/s	Central Durat.
10:37.1	36°37.2'S	050°02.3'W	1.0342	1.0695	0.0	60.3	127.4	–	115.1	–	02m06.1s
10:40	30°38.9'S	037°56.6'W	1.0380	1.0774	12.2	53.1	139.9	599.4	127.4	3.537	02m34.3s
10:45	26°56.2'S	031°05.1'W	1.0404	1.0824	20.0	48.7	148.5	392.9	135.2	2.039	02m55.4s
10:50	24°25.3'S	026°36.2'W	1.0420	1.0858	25.5	45.5	154.7	324.8	140.4	1.538	03m11.6s
10:55	22°26.6'S	023°06.8'W	1.0433	1.0884	30.0	42.7	160.0	288.4	144.5	1.268	03m25.4s
11:00	20°47.7'S	020°11.5'W	1.0443	1.0906	33.8	40.0	164.6	265.4	147.8	1.094	03m37.8s
11:05	19°22.9'S	017°38.6'W	1.0452	1.0925	37.1	37.4	168.9	249.3	150.7	0.971	03m49.1s
11:10	18°08.7'S	015°21.5'W	1.0460	1.0941	40.1	34.6	172.9	237.4	153.1	0.878	03m59.4s
11:15	17°03.3'S	013°16.1'W	1.0466	1.0955	42.7	31.8	176.7	228.4	155.2	0.807	04m08.8s
11:20	16°05.2'S	011°19.6'W	1.0472	1.0967	45.1	28.8	180.2	221.4	157.0	0.749	04m17.5s
11:25	15°13.4'S	009°30.1'W	1.0477	1.0977	47.2	25.7	183.5	215.8	158.6	0.703	04m25.2s
11:30	14°27.3'S	007°46.0'W	1.0481	1.0986	49.1	22.3	186.6	211.4	160.0	0.666	04m32.2s
11:35	13°46.3'S	006°06.1'W	1.0485	1.0994	50.7	18.7	189.5	207.9	161.1	0.635	04m38.4s
11:40	13°10.0'S	004°29.5'W	1.0488	1.1000	52.1	14.9	192.1	205.2	162.1	0.610	04m43.7s
11:45	12°38.2'S	002°55.4'W	1.0491	1.1005	53.3	10.9	194.4	203.1	162.9	0.591	04m48.1s
11:50	12°10.5'S	001°23.1'W	1.0493	1.1009	54.1	6.8	196.4	201.6	163.5	0.575	04m51.6s
11:55	11°46.9'S	000°07.9'E	1.0494	1.1012	54.7	2.5	198.1	200.6	164.0	0.564	04m54.2s
12:00	11°27.3'S	001°38.3'E	1.0495	1.1014	55.1	358.1	199.3	200.2	164.3	0.558	04m55.9s
12:05	11°11.5'S	003°08.5'E	1.0495	1.1015	55.2	353.6	200.1	200.2	164.4	0.554	04m56.6s
12:10	10°59.6'S	004°39.1'E	1.0495	1.1015	54.9	349.2	200.5	200.7	164.4	0.555	04m56.3s
12:15	10°51.5'S	006°10.5'E	1.0495	1.1014	54.5	344.9	200.3	201.6	164.2	0.560	04m55.1s
12:20	10°47.4'S	007°43.3'E	1.0494	1.1012	53.7	340.7	199.7	203.1	163.9	0.569	04m52.8s
12:25	10°47.3'S	009°18.0'E	1.0492	1.1008	52.7	336.7	198.6	205.2	163.4	0.583	04m49.6s
12:30	10°51.5'S	010°55.3'E	1.0490	1.1004	51.5	333.0	197.0	207.9	162.7	0.601	04m45.5s
12:35	11°00.0'S	012°35.8'E	1.0487	1.0998	49.9	329.4	195.0	211.3	161.8	0.625	04m40.3s
12:40	11°13.4'S	014°20.4'E	1.0484	1.0991	48.2	326.1	192.5	215.5	160.8	0.656	04m34.2s
12:45	11°31.8'S	016°10.2'E	1.0480	1.0983	46.2	323.0	189.7	220.9	159.5	0.695	04m27.2s
12:50	11°56.0'S	018°06.3'E	1.0475	1.0973	43.9	320.2	186.4	227.5	158.0	0.744	04m19.3s
12:55	12°26.6'S	020°10.4'E	1.0470	1.0961	41.4	317.5	182.8	236.0	156.2	0.806	04m10.4s
13:00	13°04.7'S	022°24.7'E	1.0463	1.0948	38.6	315.1	178.8	246.8	154.2	0.885	04m00.6s
13:05	13°51.7'S	024°52.6'E	1.0456	1.0932	35.5	312.7	174.4	261.3	151.8	0.990	03m49.7s
13:10	14°50.1'S	027°38.8'E	1.0446	1.0913	31.9	310.5	169.5	281.3	148.9	1.134	03m37.8s
13:15	16°03.5'S	030°51.5'E	1.0436	1.0890	27.8	308.3	164.1	311.1	145.4	1.348	03m24.4s
13:20	17°39.6'S	034°46.7'E	1.0422	1.0862	22.9	306.0	157.9	361.6	141.0	1.708	03m09.1s
13:25	19°56.8'S	040°04.4'E	1.0403	1.0823	16.5	303.3	150.0	473.8	135.0	2.500	02m50.6s
13:30	24°53.5'S	051°05.3'E	1.0366	1.0745	4.2	298.3	135.8	1639.1	122.9	10.643	02m19.3s
13:30.4	26°45.0'S	055°13.6'E	1.0352	1.0717	0.0	296.5	131.2	–	118.6	–	02m09.8s

TABLE 5

LOCAL CIRCUMSTANCES ON THE CENTER LINE
TOTAL SOLAR ECLIPSE OF 2001 JUNE 21

Center Line Maximum Eclipse			First Contact				Second Contact			Third Contact			Fourth Contact			.
U.T.	Durat.	Alt °	U.T.	P °	V °	Alt °	U.T.	P °	V °	U.T.	P °	V °	U.T.	P °	V °	Alt °
10:40	02m34.3s	12	09:34:55	254	18	1	10:38:43	74	205	10:41:17	254	25	11:53:36	74	216	24
10:45	02m55.4s	20	09:35:47	253	17	8	10:43:33	73	206	10:46:28	253	26	12:04:12	73	221	32
10:50	03m11.6s	26	09:37:35	253	17	13	10:48:25	73	207	10:51:36	253	28	12:13:21	74	226	37
10:55	03m25.4s	30	09:39:47	253	17	17	10:53:18	73	209	10:56:43	253	30	12:21:49	74	232	41
11:00	03m37.8s	34	09:42:14	252	17	21	10:58:11	73	212	11:01:49	253	32	12:29:46	75	237	44
11:05	03m49.1s	37	09:44:52	252	18	24	11:03:06	73	214	11:06:55	253	35	12:37:18	75	243	46
11:10	03m59.4s	40	09:47:37	253	19	27	11:08:01	74	217	11:12:00	254	38	12:44:29	76	249	48
11:15	04m08.8s	43	09:50:31	253	20	29	11:12:56	74	220	11:17:05	254	41	12:51:19	77	255	50
11:20	04m17.5s	45	09:53:30	253	22	32	11:17:52	74	223	11:22:09	254	45	12:57:51	77	261	50
11:25	04m25.2s	47	09:56:36	253	23	34	11:22:48	75	227	11:27:13	255	49	13:04:03	78	268	51
11:30	04m32.2s	49	09:59:47	254	25	37	11:27:44	76	231	11:32:16	256	53	13:09:59	79	274	51
11:35	04m38.4s	51	10:03:05	254	27	39	11:32:41	76	235	11:37:19	256	57	13:15:37	80	279	51
11:40	04m43.7s	52	10:06:28	254	29	41	11:37:38	77	240	11:42:22	257	62	13:21:00	81	285	50
11:45	04m48.1s	53	10:09:59	255	32	43	11:42:36	78	245	11:47:24	258	67	13:26:08	82	290	50
11:50	04m51.6s	54	10:13:36	255	34	45	11:47:34	78	250	11:52:26	259	72	13:31:01	83	295	49
11:55	04m54.2s	55	10:17:21	256	37	47	11:52:33	79	256	11:57:27	259	78	13:35:41	84	299	48
12:00	04m55.9s	55	10:21:15	257	41	48	11:57:32	80	261	12:02:28	260	83	13:40:09	84	304	47
12:05	04m56.6s	55	10:25:18	257	44	50	12:02:32	81	267	12:07:28	261	89	13:44:25	85	308	45
12:10	04m56.3s	55	10:29:30	258	48	51	12:07:32	82	272	12:12:28	262	95	13:48:30	86	311	44
12:15	04m55.1s	54	10:33:54	259	53	52	12:12:32	83	278	12:17:27	263	100	13:52:25	87	315	42
12:20	04m52.8s	54	10:38:29	260	58	53	12:17:33	84	284	12:22:26	264	106	13:56:11	88	318	41
12:25	04m49.6s	53	10:43:18	261	63	54	12:22:35	85	289	12:27:25	265	111	13:59:47	89	321	39
12:30	04m45.5s	51	10:48:21	262	69	55	12:27:37	86	294	12:32:22	266	116	14:03:16	89	324	37
12:35	04m40.3s	50	10:53:39	263	75	55	12:32:40	87	299	12:37:20	267	121	14:06:37	90	326	35
12:40	04m34.2s	48	10:59:14	264	81	55	12:37:43	88	304	12:42:17	268	125	14:09:51	91	328	33
12:45	04m27.2s	46	11:05:07	265	88	55	12:42:46	88	308	12:47:13	269	129	14:12:58	91	331	31
12:50	04m19.3s	44	11:11:20	266	95	54	12:47:50	89	312	12:52:09	269	133	14:15:57	92	333	28
12:55	04m10.4s	41	11:17:53	267	102	53	12:52:54	90	316	12:57:05	270	137	14:18:50	92	334	26
13:00	04m00.6s	39	11:24:49	268	109	51	12:57:59	91	319	13:02:00	271	140	14:21:35	93	336	23
13:05	03m49.7s	35	11:32:09	269	115	49	13:03:05	92	322	13:06:54	272	143	14:24:11	93	337	20
13:10	03m37.8s	32	11:39:56	270	122	46	13:08:11	92	325	13:11:49	272	146	14:26:38	94	339	17
13:15	03m24.4s	28	11:48:15	271	128	42	13:13:17	93	328	13:16:42	273	149	14:28:51	94	340	13
13:20	03m09.1s	23	11:57:12	272	134	37	13:18:25	94	331	13:21:34	274	151	14:30:45	94	341	9
13:25	02m50.6s	17	12:07:12	273	139	31	13:23:34	94	333	13:26:25	274	153	14:32:04	94	341	3
13:30	02m19.3s	4	12:20:49	274	146	17	13:28:50	94	335	13:31:09	274	155	-	-	-	-

TABLE 6

TOPOCENTRIC DATA AND PATH CORRECTIONS DUE TO LUNAR LIMB PROFILE
TOTAL SOLAR ECLIPSE OF 2001 JUNE 21

Universal Time	Moon Topo H.P. "	Moon Topo S.D. "	Moon Rel. Ang.V "/s	Topo Lib. Long °	Sun Alt. °	Sun Az. °	Path Az. °	North Limit P.A. °	North Limit Int. '	North Limit Ext. '	South Limit Int. '	South Limit Ext. '	Central Durat. Cor. s
10:40	3599.4	980.1	0.465	-3.20	12.2	53.1	58.6	343.6	-1.2	0.9	0.9	-3.6	-0.4
10:45	3607.8	982.3	0.435	-3.24	20.0	48.7	58.1	343.0	-1.2	1.1	0.8	-4.0	-0.1
10:50	3613.4	983.9	0.414	-3.28	25.5	45.5	58.1	342.9	-1.2	1.2	0.8	-4.1	0.0
10:55	3617.9	985.1	0.398	-3.32	30.0	42.7	58.5	342.9	-1.2	1.2	0.8	-4.1	0.2
11:00	3621.6	986.1	0.384	-3.37	33.8	40.0	59.1	343.0	-1.1	1.2	0.8	-4.1	0.3
11:05	3624.7	986.9	0.373	-3.41	37.1	37.4	59.8	343.2	-1.1	1.1	0.8	-3.9	0.3
11:10	3627.4	987.6	0.363	-3.45	40.1	34.6	60.8	343.5	-1.0	1.0	0.9	-3.7	0.3
11:15	3629.7	988.3	0.354	-3.50	42.7	31.8	61.9	344.0	-0.8	0.7	0.9	-3.3	0.2
11:20	3631.7	988.8	0.346	-3.54	45.1	28.8	63.2	344.4	-0.8	0.5	1.0	-2.9	0.1
11:25	3633.5	989.3	0.340	-3.58	47.2	25.7	64.6	345.0	-0.8	0.7	1.1	-2.3	0.0
11:30	3635.0	989.7	0.334	-3.62	49.1	22.3	66.2	345.6	-0.7	0.9	1.1	-1.7	-0.2
11:35	3636.3	990.0	0.329	-3.67	50.7	18.7	68.0	346.3	-0.7	0.9	1.2	-2.4	-0.5
11:40	3637.4	990.3	0.325	-3.71	52.1	14.9	69.9	347.0	-0.7	0.8	1.3	-2.9	-0.4
11:45	3638.2	990.6	0.322	-3.75	53.3	10.9	71.9	347.7	-0.6	0.6	1.3	-3.2	-0.7
11:50	3638.9	990.8	0.319	-3.80	54.1	6.8	74.0	348.5	-0.6	0.5	1.4	-3.3	-1.0
11:55	3639.4	990.9	0.317	-3.84	54.7	2.5	76.3	349.4	-0.5	0.6	1.4	-3.1	-1.3
12:00	3639.8	991.0	0.316	-3.88	55.1	358.1	78.7	350.3	-0.4	0.8	1.5	-2.8	-1.5
12:05	3639.9	991.0	0.315	-3.92	55.2	353.6	81.1	351.2	-0.3	0.8	1.5	-2.1	-1.4
12:10	3639.9	991.0	0.316	-3.97	54.9	349.2	83.6	352.1	-0.2	0.7	1.5	-1.2	-1.6
12:15	3639.7	991.0	0.317	-4.01	54.5	344.9	86.1	353.0	-0.0	0.8	1.5	-0.9	-1.8
12:20	3639.3	990.9	0.318	-4.05	53.7	340.7	88.7	354.0	-0.1	0.6	1.5	-1.3	-1.8
12:25	3638.8	990.7	0.321	-4.10	52.7	336.7	91.2	354.9	-0.2	0.7	1.4	-1.8	-1.7
12:30	3638.0	990.5	0.324	-4.14	51.5	333.0	93.7	355.8	-0.2	0.8	1.4	-1.9	-1.1
12:35	3637.0	990.2	0.328	-4.18	49.9	329.4	96.2	356.7	-0.3	0.6	1.3	-1.7	-0.9
12:40	3635.9	989.9	0.333	-4.22	48.2	326.1	98.6	357.7	-0.3	0.5	1.2	-1.6	-0.7
12:45	3634.5	989.6	0.339	-4.27	46.2	323.0	100.9	358.5	-0.3	0.6	1.2	-1.4	-1.1
12:50	3632.8	989.1	0.346	-4.31	43.9	320.2	103.1	359.4	-0.2	0.6	1.1	-1.3	-1.1
12:55	3630.8	988.6	0.354	-4.35	41.4	317.5	105.2	0.2	-0.2	0.7	1.0	-1.8	-1.9
13:00	3628.6	988.0	0.364	-4.39	38.6	315.1	107.2	1.0	-0.1	0.8	0.9	-2.0	-2.5
13:05	3625.9	987.2	0.374	-4.44	35.5	312.7	109.0	1.7	-0.1	0.7	0.8	-2.0	-3.1
13:10	3622.7	986.4	0.387	-4.48	31.9	310.5	110.7	2.4	-0.1	0.6	0.7	-2.1	-3.4
13:15	3618.9	985.4	0.402	-4.52	27.8	308.3	112.3	3.0	0.0	0.7	0.6	-2.5	-3.7
13:20	3614.1	984.1	0.421	-4.57	22.9	306.0	113.8	3.5	0.1	0.9	0.6	-2.7	-3.9
13:25	3607.5	982.3	0.446	-4.61	16.5	303.3	115.1	4.0	0.1	0.9	0.6	-2.8	-4.0
13:30	3594.6	978.7	0.496	-4.65	4.2	298.3	116.5	4.1	0.1	0.9	0.6	-2.9	-3.9

TABLE 7
MAPPING COORDINATES FOR THE UMBRAL PATH
TOTAL SOLAR ECLIPSE OF 2001 JUNE 21

Longitude	Latitude of: Northern Limit	Southern Limit	Center Line	Circumstances on Center Line Universal Time h m s	Sun Alt °	Sun Az. °	Path Width km	Central Durat.
050°00.0'W	35°56.56'S	—	36°36.16'S	10:37:05	0	60	127	02m06.1s
049°00.0'W	35°28.42'S	36°48.46'S	36°08.37'S	10:37:06	1	60	128	02m08.1s
048°00.0'W	34°59.93'S	36°20.70'S	35°40.25'S	10:37:09	2	59	129	02m10.1s
047°00.0'W	34°31.10'S	35°52.60'S	35°11.79'S	10:37:15	3	58	130	02m12.2s
046°00.0'W	34°01.92'S	35°24.17'S	34°42.98'S	10:37:22	4	58	131	02m14.4s
045°00.0'W	33°32.40'S	34°55.40'S	34°13.84'S	10:37:32	5	57	132	02m16.6s
044°00.0'W	33°02.54'S	34°26.31'S	33°44.37'S	10:37:45	6	57	133	02m18.9s
043°00.0'W	32°32.34'S	33°56.89'S	33°14.56'S	10:38:00	7	56	134	02m21.3s
042°00.0'W	32°01.81'S	33°27.15'S	32°44.43'S	10:38:18	8	56	135	02m23.7s
041°00.0'W	31°30.95'S	32°57.10'S	32°13.97'S	10:38:38	9	55	136	02m26.2s
040°00.0'W	30°59.76'S	32°26.73'S	31°43.20'S	10:39:02	10	54	138	02m28.8s
039°00.0'W	30°28.26'S	31°56.05'S	31°12.11'S	10:39:29	11	54	139	02m31.4s
038°00.0'W	29°56.46'S	31°25.07'S	30°40.72'S	10:39:58	12	53	140	02m34.2s
037°00.0'W	29°24.35'S	30°53.81'S	30°09.04'S	10:40:31	13	52	141	02m37.0s
036°00.0'W	28°51.94'S	30°22.26'S	29°37.07'S	10:41:08	14	52	142	02m39.9s
035°00.0'W	28°19.26'S	29°50.44'S	29°04.82'S	10:41:47	15	51	143	02m42.9s
034°00.0'W	27°46.32'S	29°18.36'S	28°32.31'S	10:42:31	17	51	145	02m45.9s
033°00.0'W	27°13.12'S	28°46.03'S	27°59.54'S	10:43:18	18	50	146	02m49.1s
032°00.0'W	26°39.68'S	28°13.47'S	27°26.55'S	10:44:10	19	49	147	02m52.4s
031°00.0'W	26°06.03'S	27°40.70'S	26°53.34'S	10:45:05	20	49	149	02m55.7s
030°00.0'W	25°32.18'S	27°07.74'S	26°19.93'S	10:46:04	21	48	150	02m59.2s
029°00.0'W	24°58.16'S	26°34.60'S	25°46.35'S	10:47:08	23	47	151	03m02.7s
028°00.0'W	24°23.98'S	26°01.31'S	25°12.62'S	10:48:17	24	47	153	03m06.3s
027°00.0'W	23°49.69'S	25°27.89'S	24°38.77'S	10:49:30	25	46	154	03m10.1s
026°00.0'W	23°15.30'S	24°54.38'S	24°04.82'S	10:50:48	26	45	156	03m13.9s
025°00.0'W	22°40.86'S	24°20.81'S	23°30.82'S	10:52:10	28	44	157	03m17.8s
024°00.0'W	22°06.41'S	23°47.21'S	22°56.79'S	10:53:38	29	43	159	03m21.8s
023°00.0'W	21°31.97'S	23°13.63'S	22°22.78'S	10:55:11	30	43	160	03m25.9s
022°00.0'W	20°57.61'S	22°40.09'S	21°48.83'S	10:56:49	31	42	162	03m30.1s
021°00.0'W	20°23.36'S	22°06.65'S	21°14.99'S	10:58:32	33	41	163	03m34.3s
020°00.0'W	19°49.29'S	21°33.36'S	20°41.30'S	11:00:21	34	40	165	03m38.7s
019°00.0'W	19°15.44'S	21°00.27'S	20°07.83'S	11:02:16	35	39	167	03m43.0s
018°00.0'W	18°41.87'S	20°27.44'S	19°34.63'S	11:04:16	37	38	168	03m47.5s
017°00.0'W	18°08.67'S	19°54.93'S	19°01.77'S	11:06:21	38	37	170	03m52.0s
016°00.0'W	17°35.88'S	19°22.81'S	18°29.31'S	11:08:33	39	35	172	03m56.5s
015°00.0'W	17°03.59'S	18°51.14'S	17°57.33'S	11:10:50	41	34	174	04m01.0s
014°00.0'W	16°31.87'S	18°19.99'S	17°25.89'S	11:13:12	42	33	175	04m05.5s
013°00.0'W	16°00.80'S	17°49.45'S	16°55.08'S	11:15:40	43	31	177	04m10.0s
012°00.0'W	15°30.46'S	17°19.59'S	16°24.98'S	11:18:14	44	30	179	04m14.5s
011°00.0'W	15°00.94'S	16°50.50'S	15°55.67'S	11:20:53	46	28	181	04m18.9s

Longitude	Latitude of:			Circumstances on Center Line				
	Northern Limit	Southern Limit	Center Line	Universal Time h m s	Sun Alt °	Sun Az. °	Path Width km	Central Durat.
010°00.0'W	14°32.32'S	16°22.25'S	15°27.23'S	11:23:36	47	27	183	04m23.2s
009°00.0'W	14°04.69'S	15°54.94'S	14°59.75'S	11:26:25	48	25	184	04m27.3s
008°00.0'W	13°38.13'S	15°28.64'S	14°33.31'S	11:29:19	49	23	186	04m31.3s
007°00.0'W	13°12.73'S	15°03.44'S	14°08.00'S	11:32:17	50	21	188	04m35.1s
006°00.0'W	12°48.57'S	14°39.42'S	13°43.91'S	11:35:19	51	18	190	04m38.7s
005°00.0'W	12°25.72'S	14°16.66'S	13°21.10'S	11:38:24	52	16	191	04m42.1s
004°00.0'W	12°04.27'S	13°55.25'S	12°59.66'S	11:41:33	53	14	193	04m45.2s
003°00.0'W	11°44.27'S	13°35.24'S	12°39.65'S	11:44:45	53	11	194	04m47.9s
002°00.0'W	11°25.80'S	13°16.70'S	12°21.14'S	11:47:59	54	8	196	04m50.3s
001°00.0'W	11°08.91'S	12°59.70'S	12°04.19'S	11:51:16	54	6	197	04m52.4s
000°00.0'-	10°53.64'S	12°44.28'S	11°48.84'S	11:54:34	55	3	198	04m54.1s
001°00.0'E	10°40.04'S	12°30.50'S	11°35.14'S	11:57:53	55	360	199	04m55.3s
002°00.0'E	10°28.14'S	12°18.37'S	11°23.13'S	12:01:12	55	357	200	04m56.2s
003°00.0'E	10°17.96'S	12°07.93'S	11°12.82'S	12:04:32	55	354	200	04m56.6s
004°00.0'E	10°09.51'S	11°59.20'S	11°04.23'S	12:07:51	55	351	200	04m56.6s
005°00.0'E	10°02.80'S	11°52.19'S	10°57.36'S	12:11:09	55	348	200	04m56.1s
006°00.0'E	09°57.83'S	11°46.89'S	10°52.22'S	12:14:26	55	345	200	04m55.3s
007°00.0'E	09°54.58'S	11°43.29'S	10°48.80'S	12:17:41	54	343	200	04m54.0s
008°00.0'E	09°53.03'S	11°41.38'S	10°47.07'S	12:20:54	54	340	200	04m52.3s
009°00.0'E	09°53.17'S	11°41.13'S	10°47.01'S	12:24:04	53	337	199	04m50.3s
010°00.0'E	09°54.94'S	11°42.51'S	10°48.59'S	12:27:11	52	335	198	04m47.9s
011°00.0'E	09°58.32'S	11°45.47'S	10°51.77'S	12:30:14	51	333	197	04m45.2s
012°00.0'E	10°03.26'S	11°49.99'S	10°56.50'S	12:33:14	51	331	196	04m42.2s
013°00.0'E	10°09.72'S	11°56.01'S	11°02.74'S	12:36:11	50	329	194	04m39.0s
014°00.0'E	10°17.63'S	12°03.47'S	11°10.43'S	12:39:02	49	327	193	04m35.5s
015°00.0'E	10°26.95'S	12°12.32'S	11°19.52'S	12:41:50	47	325	192	04m31.8s
016°00.0'E	10°37.62'S	12°22.51'S	11°29.95'S	12:44:33	46	323	190	04m27.9s
017°00.0'E	10°49.58'S	12°33.98'S	11°41.67'S	12:47:11	45	322	188	04m23.9s
018°00.0'E	11°02.76'S	12°46.65'S	11°54.60'S	12:49:44	44	320	187	04m19.7s
019°00.0'E	11°17.11'S	13°00.48'S	12°08.70'S	12:52:13	43	319	185	04m15.5s
020°00.0'E	11°32.57'S	13°15.40'S	12°23.89'S	12:54:36	42	318	183	04m11.2s
021°00.0'E	11°49.07'S	13°31.36'S	12°40.12'S	12:56:54	40	317	181	04m06.8s
022°00.0'E	12°06.55'S	13°48.28'S	12°57.32'S	12:59:07	39	316	180	04m02.4s
023°00.0'E	12°24.96'S	14°06.11'S	13°15.45'S	13:01:14	38	314	178	03m58.0s
024°00.0'E	12°44.24'S	14°24.80'S	13°34.43'S	13:03:17	37	314	176	03m53.6s
025°00.0'E	13°04.33'S	14°44.28'S	13°54.22'S	13:05:14	35	313	174	03m49.2s

TABLE 7 - continued

MAPPING COORDINATES FOR THE UMBRAL PATH
TOTAL SOLAR ECLIPSE OF 2001 JUNE 21

Longitude	Latitude of:			Circumstances on Center Line				
	Northern Limit	Southern Limit	Center Line	Universal Time h m s	Sun Alt °	Sun Az. °	Path Width km	Central Durat.
026°00.0'E	13°25.17'S	15°04.50'S	14°14.76'S	13:07:06	34	312	172	03m44.8s
027°00.0'E	13°46.71'S	15°25.42'S	14°35.99'S	13:08:53	33	311	171	03m40.5s
028°00.0'E	14°08.90'S	15°46.97'S	14°57.86'S	13:10:36	31	310	169	03m36.3s
029°00.0'E	14°31.70'S	16°09.11'S	15°20.34'S	13:12:13	30	310	167	03m32.1s
030°00.0'E	14°55.04'S	16°31.80'S	15°43.35'S	13:13:45	29	309	166	03m27.9s
031°00.0'E	15°18.90'S	16°54.99'S	16°06.88'S	13:15:12	28	308	164	03m23.8s
032°00.0'E	15°43.22'S	17°18.63'S	16°30.86'S	13:16:35	26	308	162	03m19.8s
033°00.0'E	16°07.96'S	17°42.69'S	16°55.26'S	13:17:53	25	307	161	03m15.9s
034°00.0'E	16°33.09'S	18°07.13'S	17°20.05'S	13:19:06	24	306	159	03m12.1s
035°00.0'E	16°58.57'S	18°31.91'S	17°45.18'S	13:20:15	23	306	158	03m08.3s
036°00.0'E	17°24.36'S	18°57.00'S	18°10.62'S	13:21:20	21	305	156	03m04.7s
037°00.0'E	17°50.42'S	19°22.37'S	18°36.34'S	13:22:20	20	305	154	03m01.1s
038°00.0'E	18°16.74'S	19°47.98'S	19°02.30'S	13:23:16	19	304	153	02m57.6s
039°00.0'E	18°43.27'S	20°13.81'S	19°28.48'S	13:24:08	18	304	152	02m54.2s
040°00.0'E	19°10.00'S	20°39.83'S	19°54.85'S	13:24:57	17	303	150	02m50.8s
041°00.0'E	19°36.89'S	21°06.01'S	20°21.39'S	13:25:41	15	303	149	02m47.6s
042°00.0'E	20°03.92'S	21°32.33'S	20°48.06'S	13:26:22	14	302	147	02m44.4s
043°00.0'E	20°31.06'S	21°58.77'S	21°14.86'S	13:26:59	13	302	146	02m41.3s
044°00.0'E	20°58.30'S	22°25.30'S	21°41.74'S	13:27:33	12	301	145	02m38.4s
045°00.0'E	21°25.62'S	22°51.92'S	22°08.71'S	13:28:03	11	301	143	02m35.4s
046°00.0'E	21°52.99'S	23°18.59'S	22°35.73'S	13:28:30	10	301	142	02m32.6s
047°00.0'E	22°20.39'S	23°45.30'S	23°02.78'S	13:28:54	9	300	141	02m29.8s
048°00.0'E	22°47.82'S	24°12.04'S	23°29.87'S	13:29:14	8	300	140	02m27.2s
049°00.0'E	23°15.25'S	24°38.78'S	23°56.95'S	13:29:32	6	299	138	02m24.5s
050°00.0'E	23°42.68'S	25°05.52'S	24°24.03'S	13:29:47	5	299	137	02m22.0s
051°00.0'E	24°10.07'S	25°32.24'S	24°51.09'S	13:29:59	4	298	136	02m19.5s
052°00.0'E	24°37.44'S	25°58.94'S	25°18.12'S	13:30:08	3	298	135	02m17.1s
053°00.0'E	25°04.75'S	26°25.58'S	25°45.10'S	13:30:15	2	297	134	02m14.8s
054°00.0'E	25°32.00'S	26°52.18'S	26°12.02'S	13:30:19	1	297	133	02m12.5s

TABLE 8

MAPPING COORDINATES FOR THE ZONES OF GRAZING ECLIPSE
TOTAL SOLAR ECLIPSE OF 2001 JUNE 21

Longitude	North Graze Zone Latitudes		Northern Limit	South Graze Zone Latitudes		Southern Limit .	Path Azm	Elev Fact	Scale Fact
	Northern Limit	Southern Limit	Universal Time	Northern Limit	Southern Limit	Universal Time			
° ′	° ′	° ′	h m s	° ′	° ′	h m s	°		km/″
013 00.0E	10 09.13S	10 09.97S	12:36:29	11 54.70S	11 57.62S	12:35:51	96.8	0.67	2.14
013 30.0E	10 12.93S	10 13.76S	12:37:56	11 58.28S	12 01.14S	12:37:17	97.4	0.67	2.14
014 00.0E	10 17.10S	10 17.90S	12:39:21	12 02.21S	12 05.06S	12:38:43	98.1	0.66	2.13
014 30.0E	10 21.63S	10 22.39S	12:40:45	12 06.49S	12 09.29S	12:40:07	98.8	0.66	2.13
015 00.0E	10 26.39S	10 27.22S	12:42:08	12 11.11S	12 13.86S	12:41:31	99.4	0.65	2.12
015 30.0E	10 31.51S	10 32.39S	12:43:30	12 16.06S	12 18.75S	12:42:53	100.1	0.65	2.12
016 00.0E	10 36.98S	10 37.89S	12:44:50	12 21.34S	12 23.95S	12:44:14	100.7	0.65	2.12
016 30.0E	10 42.79S	10 43.70S	12:46:10	12 26.94S	12 29.44S	12:45:34	101.3	0.64	2.11
017 00.0E	10 48.93S	10 49.83S	12:47:28	12 32.85S	12 35.23S	12:46:52	101.9	0.64	2.11
017 30.0E	10 55.39S	10 56.27S	12:48:45	12 39.06S	12 41.31S	12:48:10	102.4	0.63	2.10
018 00.0E	11 02.12S	11 02.96S	12:50:01	12 45.57S	12 47.89S	12:49:26	103.0	0.63	2.10
018 30.0E	11 09.21S	11 09.99S	12:51:16	12 52.37S	12 54.81S	12:50:41	103.5	0.62	2.09
019 00.0E	11 16.54S	11 17.30S	12:52:29	12 59.44S	13 01.99S	12:51:54	104.0	0.62	2.09
019 30.0E	11 24.06S	11 24.89S	12:53:42	13 06.80S	13 09.43S	12:53:06	104.5	0.61	2.08
020 00.0E	11 31.86S	11 32.75S	12:54:53	13 14.41S	13 17.11S	12:54:17	105.0	0.61	2.08
020 30.0E	11 39.94S	11 40.86S	12:56:02	13 22.28S	13 25.04S	12:55:27	105.5	0.60	2.08
021 00.0E	11 48.28S	11 49.24S	12:57:11	13 30.40S	13 33.20S	12:56:35	105.9	0.60	2.07
021 30.0E	11 56.89S	11 57.85S	12:58:18	13 38.77S	13 41.59S	12:57:42	106.4	0.59	2.07
022 00.0E	12 05.74S	12 06.71S	12:59:24	13 47.37S	13 50.21S	12:58:48	106.8	0.59	2.06
022 30.0E	12 14.83S	12 15.79S	13:00:28	13 56.19S	13 59.04S	12:59:52	107.2	0.59	2.06
023 00.0E	12 24.16S	12 25.10S	13:01:31	14 05.24S	14 08.09S	13:00:56	107.6	0.58	2.06
023 30.0E	12 33.71S	12 34.63S	13:02:33	14 14.50S	14 17.34S	13:01:57	108.0	0.58	2.05
024 00.0E	12 43.48S	12 44.36S	13:03:34	14 23.96S	14 26.78S	13:02:58	108.4	0.57	2.05
024 30.0E	12 53.45S	12 54.30S	13:04:34	14 33.62S	14 36.42S	13:03:57	108.8	0.57	2.04
025 00.0E	13 03.63S	13 04.43S	13:05:32	14 43.48S	14 46.25S	13:04:55	109.1	0.56	2.04
025 30.0E	13 14.00S	13 14.76S	13:06:29	14 53.52S	14 56.25S	13:05:51	109.4	0.56	2.04
026 00.0E	13 24.55S	13 25.26S	13:07:24	15 03.74S	15 06.43S	13:06:47	109.8	0.56	2.03
026 30.0E	13 35.28S	13 35.94S	13:08:19	15 14.13S	15 16.78S	13:07:41	110.1	0.55	2.03
027 00.0E	13 46.16S	13 46.79S	13:09:12	15 24.68S	15 27.34S	13:08:33	110.4	0.55	2.03
027 30.0E	13 57.17S	13 57.80S	13:10:04	15 35.39S	15 38.14S	13:09:25	110.7	0.54	2.02
028 00.0E	14 08.32S	14 08.97S	13:10:54	15 46.26S	15 49.07S	13:10:15	110.9	0.54	2.02
028 30.0E	14 19.62S	14 20.29S	13:11:43	15 57.27S	16 00.15S	13:11:04	111.2	0.54	2.02
029 00.0E	14 31.07S	14 31.73S	13:12:32	16 08.43S	16 11.37S	13:11:52	111.5	0.53	2.01
029 30.0E	14 42.64S	14 43.29S	13:13:19	16 19.72S	16 22.70S	13:12:38	111.7	0.53	2.01
030 00.0E	14 54.34S	14 55.02S	13:14:04	16 31.14S	16 34.16S	13:13:23	111.9	0.53	2.01
030 30.0E	15 06.18S	15 06.88S	13:14:49	16 42.68S	16 45.75S	13:14:07	112.2	0.52	2.01
031 00.0E	15 18.14S	15 18.87S	13:15:32	16 54.35S	16 57.45S	13:14:50	112.4	0.52	2.00
031 30.0E	15 30.22S	15 30.97S	13:16:14	17 06.12S	17 09.26S	13:15:32	112.6	0.52	2.00
032 00.0E	15 42.41S	15 43.18S	13:16:55	17 18.01S	17 21.18S	13:16:12	112.8	0.51	2.00
032 30.0E	15 54.71S	15 55.49S	13:17:35	17 30.00S	17 33.20S	13:16:52	113.0	0.51	2.00
033 00.0E	16 07.12S	16 07.91S	13:18:14	17 42.08S	17 45.31S	13:17:30	113.2	0.51	1.99
033 30.0E	16 19.62S	16 20.43S	13:18:51	17 54.26S	17 57.51S	13:18:07	113.4	0.51	1.99

TABLE 8 - continued

MAPPING COORDINATES FOR THE ZONES OF GRAZING ECLIPSE
TOTAL SOLAR ECLIPSE OF 1999 AUGUST 11

Longitude	North Graze Zone Latitudes		Northern Limit	South Graze Zone Latitudes		Southern Limit .	Path	Elev	Scale
	Northern Limit	Southern Limit	Universal Time	Northern Limit	Southern Limit	Universal Time	Azm	Fact	Fact
° ′	° ′	° ′	h m s	° ′	° ′	h m s	°		km/″
034 00.0E	16 32.22S	16 33.03S	13:19:27	18 06.53S	18 09.80S	13:18:43	113.5	0.50	1.99
034 30.0E	16 44.91S	16 45.72S	13:20:03	18 18.89S	18 22.18S	13:19:18	113.7	0.50	1.99
035 00.0E	16 57.68S	16 58.50S	13:20:37	18 31.32S	18 34.63S	13:19:51	113.9	0.50	1.99
035 30.0E	17 10.53S	17 11.35S	13:21:10	18 43.84S	18 47.16S	13:20:24	114.0	0.50	1.98
036 00.0E	17 23.46S	17 24.28S	13:21:42	18 56.42S	18 59.76S	13:20:55	114.2	0.49	1.98
036 30.0E	17 36.45S	17 37.28S	13:22:13	19 09.08S	19 12.42S	13:21:26	114.3	0.49	1.98
037 00.0E	17 49.52S	17 50.35S	13:22:43	19 21.79S	19 25.15S	13:21:55	114.4	0.49	1.98
037 30.0E	18 02.64S	18 03.47S	13:23:12	19 34.57S	19 37.94S	13:22:24	114.6	0.49	1.98
038 00.0E	18 15.83S	18 16.65S	13:23:40	19 47.41S	19 50.79S	13:22:51	114.7	0.49	1.98
038 30.0E	18 29.07S	18 29.90S	13:24:06	20 00.30S	20 03.68S	13:23:17	114.8	0.48	1.97
039 00.0E	18 42.36S	18 43.19S	13:24:32	20 13.24S	20 16.63S	13:23:43	114.9	0.48	1.97
039 30.0E	18 55.70S	18 56.53S	13:24:57	20 26.23S	20 29.63S	13:24:07	115.0	0.48	1.97
040 00.0E	19 09.08S	19 09.91S	13:25:21	20 39.26S	20 42.67S	13:24:30	115.1	0.48	1.97
040 30.0E	19 22.51S	19 23.33S	13:25:44	20 52.33S	20 55.74S	13:24:53	115.2	0.48	1.97
041 00.0E	19 35.97S	19 36.79S	13:26:06	21 05.44S	21 08.86S	13:25:14	115.3	0.48	1.97
041 30.0E	19 49.47S	19 50.29S	13:26:27	21 18.58S	21 22.01S	13:25:35	115.4	0.47	1.97
042 00.0E	20 03.00S	20 03.82S	13:26:47	21 31.76S	21 35.19S	13:25:55	115.5	0.47	1.96
042 30.0E	20 16.56S	20 17.38S	13:27:07	21 44.96S	21 48.40S	13:26:14	115.6	0.47	1.96
043 00.0E	20 30.15S	20 30.97S	13:27:25	21 58.20S	22 01.64S	13:26:31	115.6	0.47	1.96
043 30.0E	20 43.76S	20 44.58S	13:27:42	22 11.45S	22 14.90S	13:26:49	115.7	0.47	1.96
044 00.0E	20 57.39S	20 58.21S	13:27:59	22 24.73S	22 28.18S	13:27:05	115.8	0.47	1.96
044 30.0E	21 11.04S	21 11.86S	13:28:15	22 38.02S	22 41.48S	13:27:20	115.8	0.47	1.96
045 00.0E	21 24.71S	21 25.52S	13:28:30	22 51.33S	22 54.79S	13:27:35	115.9	0.46	1.96
045 30.0E	21 38.39S	21 39.21S	13:28:44	23 04.66S	23 08.12S	13:27:48	116.0	0.46	1.96
046 00.0E	21 52.08S	21 52.90S	13:28:57	23 18.00S	23 21.47S	13:28:01	116.0	0.46	1.96
046 30.0E	22 05.78S	22 06.60S	13:29:10	23 31.35S	23 34.82S	13:28:13	116.1	0.46	1.96
047 00.0E	22 19.48S	22 20.31S	13:29:21	23 44.71S	23 48.18S	13:28:25	116.1	0.46	1.95
047 30.0E	22 33.19S	22 34.02S	13:29:32	23 58.07S	24 01.55S	13:28:35	116.2	0.46	1.95
048 00.0E	22 46.91S	22 47.74S	13:29:42	24 11.43S	24 14.91S	13:28:45	116.2	0.46	1.95
048 30.0E	23 00.63S	23 01.46S	13:29:52	24 24.80S	24 28.29S	13:28:54	116.3	0.45	1.95
049 00.0E	23 14.35S	23 15.18S	13:30:01	24 38.17S	24 41.66S	13:29:02	116.3	0.45	1.95
049 30.0E	23 28.06S	23 28.89S	13:30:08	24 51.54S	24 55.03S	13:29:10	116.3	0.45	1.95
050 00.0E	23 41.78S	23 42.61S	13:30:16	25 04.90S	25 08.40S	13:29:17	116.4	0.44	1.94
050 30.0E	23 54.93S	23 54.93S	13:30:22	25 18.26S	25 21.76S	13:29:23	116.4	0.44	1.94
051 00.0E	24 10.07S	24 10.07S	13:30:28	25 31.62S	25 35.11S	13:29:29	116.5	0.43	1.94
051 30.0E	24 23.77S	24 23.77S	13:30:33	25 44.96S	25 48.46S	13:29:34	116.5	0.43	1.94
052 00.0E	24 37.45S	24 37.45S	13:30:38	25 58.29S	26 01.79S	13:29:38	116.5	0.43	1.94
052 30.0E	24 51.11S	24 51.11S	13:30:41	26 11.62S	26 15.12S	13:29:41	116.5	0.43	1.93
053 00.0E	25 04.76S	25 04.76S	13:30:45	26 24.94S	26 28.46S	13:29:44	116.5	0.43	1.93
053 30.0E	25 18.40S	25 18.40S	13:30:47	26 38.27S	26 41.75S	13:29:47	116.5	0.42	1.93
054 00.0E	25 32.01S	25 32.01S	13:30:49	26 51.65S	26 54.85S	13:29:53	116.5	0.41	1.92
054 30.0E	25 45.61S	25 45.61S	13:30:50	27 05.09S	27 07.40S	13:30:10	116.5	0.39	1.91

TABLE 9
LOCAL CIRCUMSTANCES FOR SOUTH AMERICA & SOUTH ATLANTIC
TOTAL SOLAR ECLIPSE OF 2001 JUNE 21

Location Name	Latitude	Longitude	Elev. m	First Contact U.T. h m s	P°	V°	Alt°	Second Contact U.T. h m s	P°	V°	Third Contact U.T. h m s	P°	V°	Fourth Contact U.T. h m s	P°	V°	Alt°	Maximum Eclipse U.T. h m s	P°	V°	Alt°	Azm°	Eclip. Mag.	Eclip. Obs.
ARGENTINA																								
Buenos Aires	34°36'S	058°27'W	27	—				—			—			11:35:02.6	83	215	5	11:01 Rise	—	—	0	62	0.536	0.435
Córdoba	31°24'S	064°11'W	—	—				—			—			11:28:14.9	92	218	5	11:17 Rise	—	—	0	63	0.180	0.090
General Sarmien...	34°33'S	058°43'W	—	—				—			—			11:34:49.4	84	215	5	11:03 Rise	—	—	0	62	0.503	0.398
La Plata	34°55'S	057°57'W	—	—				—			—			11:35:40.2	83	214	6	11:01 Rise	—	—	0	61	0.549	0.450
Lomas de Zamora	34°46'S	058°24'W	—	—				—			—			11:35:14.0	83	215	5	11:02 Rise	—	—	0	62	0.520	0.417
Mar del Plata	38°00'S	057°33'W	—	—				—			—			11:38:41.7	78	213	5	11:08 Rise	—	—	0	60	0.494	0.389
Morón	34°39'S	058°37'W	—	—				—			—			11:34:58.9	83	215	5	11:03 Rise	—	—	0	62	0.507	0.403
Quilmes	34°44'S	058°16'W	—	—				—			—			11:35:17.4	83	215	6	11:01 Rise	—	—	0	62	0.530	0.429
Rosario	32°57'S	060°40'W	—	—				—			—			11:32:00.4	87	216	4	11:06 Rise	—	—	0	62	0.398	0.285
San Justo	34°40'S	058°33'W	—	—				—			—			11:35:02.5	83	215	5	11:02 Rise	—	—	0	62	0.512	0.408
San Miguel de T...	26°49'S	065°13'W	—	—				—			—			11:22:08.4	101	221	2	11:10 Rise	—	—	0	64	0.172	0.084
BOLIVIA																								
Cochabamba	17°24'S	066°09'W	—	—				—			—			11:06:38.0	122	231	2	10:55 Rise	—	—	0	66	0.123	0.051
Santa Cruz	17°48'S	063°10'W	—	—				—			—			11:10:19.8	117	229	5	10:44 Rise	—	—	0	65	0.260	0.154
BRAZIL																								
Belém	01°27'S	048°29'W	13	10:17:46.3	178	275	14	—			—			10:52:06.7	147	248	21	10:34:46.7	162	262	18	65	0.036	0.008
Belford Roxo	22°46'S	043°24'W	—	09:33:25.7	237	351	0	—			—			11:41:15.0	89	220	24	10:33:29.2	163	284	11	58	0.730	0.668
Belo Horizonte	19°55'S	043°56'W	—	09:34:39.3	232	343	1	—			—			11:38:11.3	94	222	25	10:32:46.0	163	281	12	59	0.642	0.560
Brasília	15°47'S	047°55'W	1061	09:39:22.8	220	327	0	—			—			11:27:24.7	105	222	22	10:30:35.9	163	275	11	62	0.464	0.356
Campinas	22°54'S	047°05'W	—	—				—			—			11:35:56.3	92	220	20	10:31:34.2	163	282	8	60	0.681	0.608
Cava	22°41'S	043°26'W	—	09:33:27.3	237	351	0	—			—			11:41:08.3	89	220	24	10:33:27.1	163	284	11	58	0.727	0.664
Curitiba	25°25'S	049°15'W	—	—				—			—			11:35:14.5	90	218	17	10:31:20.4	164	284	5	61	0.723	0.659
Fortaleza	03°43'S	038°30'W	—	09:55:32.4	204	305	17	—			—			11:30:28.6	120	234	37	10:40:47.0	161	269	27	61	0.259	0.153
Goiânia	16°40'S	049°16'W	—	—				—			—			11:26:20.3	105	221	21	10:30:00.2	163	275	9	62	0.470	0.362
Icoraci	01°18'S	048°28'W	—	10:18:48.4	177	275	14	—			—			10:51:11.6	148	249	21	10:34:52.2	162	262	18	65	0.032	0.007
Japeri	22°39'S	043°40'W	—	09:33:25.5	237	351	0	—			—			11:40:45.2	90	220	24	10:33:18.4	163	284	11	59	0.723	0.659
Mesquita	22°48'S	043°26'W	—	09:33:26.4	237	351	0	—			—			11:41:13.4	89	220	24	10:33:28.4	163	284	11	58	0.730	0.668
Nova Iguaçu	22°45'S	043°27'W	—	—				—			—			11:41:09.7	89	220	24	10:33:27.2	163	284	11	58	0.729	0.666
Pinheirinhos	22°26'S	044°59'W	—	—				—			—			11:38:35.2	91	220	23	10:32:31.4	163	283	10	59	0.698	0.628
Porto Alegre	30°04'S	051°11'W	10	—				—			—			11:37:02.0	85	216	13	10:32:35.3	164	288	2	62	0.827	0.790
Queimados	22°42'S	043°34'W	—	—				—			—			11:40:56.7	85	220	24	10:33:22.5	163	284	11	58	0.726	0.663
Recife	08°03'S	034°54'W	30	09:45:15.5	217	323	16	—			—			11:46:34.2	106	231	41	10:42:13.4	161	275	28	58	0.442	0.332
Rio de Janeiro	22°54'S	043°14'W	61	09:39:22.4	237	352	0	—			—			11:41:36.2	89	220	24	10:33:36.8	163	284	12	58	0.736	0.675
Salvador	12°59'S	038°31'W	47	09:39:13.8	223	332	9	—			—			11:42:35.7	101	227	34	10:37:07.2	162	277	21	58	0.526	0.424
Santos	23°57'S	046°20'W	—	—				—			—			11:37:50.9	90	219	21	10:32:10.3	163	284	8	60	0.722	0.657
São Paulo	23°32'S	046°37'W	801	—				—			—			11:37:06.3	91	219	21	10:31:55.2	163	283	8	60	0.706	0.638
PARAGUAY																								
Asuncion	25°16'S	057°40'W	139	—				—			—			11:26:08.2	97	220	9	10:35 Rise	—	—	0	64	0.597	0.506
URUGUAY																								
Montevideo	34°53'S	056°11'W	22	—				—			—			11:36:51.6	81	214	7	10:53 Rise	—	—	0	62	0.688	0.616
Rivera	30°54'S	055°31'W	—	—				—			—			11:33:47.2	87	216	9	10:41 Rise	—	—	0	63	0.750	0.692
SOUTH ATLANTIC OCEAN																								
Ascension Is.	07°57'S	014°22'W	17	09:54:06.0	237	356	35	—			—			12:50:16.5	91	266	59	11:16:22.9	163	301	50	38	0.764	0.712
Bouvet Island	54°26'S	003°24'E	—	11:01:57.5	310	123	12	—			—			12:45:54.9	35	224	11	11:53:52.3	353	173	12	359	0.255	0.150
Gough Island	40°20'S	010°00'W	—	10:15:34.0	288	80	18	—			—			12:38:51.9	48	227	26	11:24:51.1	348	152	24	19	0.486	0.380
St. Helena	15°55'S	005°42'W	—	10:03:31.7	257	33	38	—			—			13:15:23.6	77	276	49	11:35:16.2	347	148	49	17	0.964	0.965
S. Georgia Is.	54°15'S	036°45'E	—	11:49:35.4	311	151	7	—			—			—				12:42:14.7	359	204	3	318	0.319	0.208
S. Sandwich Is.	57°45'S	026°30'E	—	11:37:30.4	315	146	7	—			—			13:12:47.8	38	240	1	12:25:57.4	357	194	5	330	0.249	0.144
Tristan da Cunha	37°15'S	012°30'W	—	10:06:24.8	282	69	18	—			—			12:36:49.4	52	228	29	11:18:28.2	347	146	25	24	0.562	0.466

64

TABLE 10
LOCAL CIRCUMSTANCES FOR AFRICA
TOTAL SOLAR ECLIPSE OF 2001 JUNE 21

Location Name	Latitude	Longitude	Elev. (m)	First Contact U.T. (h m s)	P (°)	V (°)	Alt	Second Contact U.T. (h m s)	P (°)	V (°)	Third Contact U.T. (h m s)	P (°)	V (°)	Fourth Contact U.T. (h m s)	P (°)	V (°)	Alt	Maximum Eclipse U.T. (h m s)	P (°)	V (°)	Alt	Azm	Eclip. Mag.	Eclip. Obs.
BENIN																								
Cotonou	06°21'N	002°26'E	--	10:46:20.0	227	2	67	--			--			13:34:29.5	115	355	60	12:10:45.9	171	7	72	346	0.504	0.402
Porto-Novo	06°29'N	002°37'E	--	10:47:04.4	227	3	67	--			--			13:34:45.4	116	355	60	12:11:19.9	171	8	72	345	0.501	0.398
BOTSWANA																								
Gaborone	24°45'S	025°55'E	--	11:34:17.0	284	126	38							14:16:23.1	77	311	15	13:01:00.3	1	223	28	317	0.753	0.697
BURKINA FASO																								
Bobo Dioulasso	11°12'N	004°18'W	--	10:48:00.4	212	327	65	--			--			13:04:41.8	125	349	74	11:55:16.0	168	322	77	24	0.309	0.199
Ouagadougou	12°22'N	001°31'W	--	10:56:36.4	211	332	70	--			--			13:10:18.0	128	4	71	12:02:59.2	170	343	79	6	0.292	0.183
BURUNDI																								
Bujumbura	03°23'S	029°22'E	--	11:52:16.5	251	119	52							14:25:58.7	114	11	21	13:15:04.7	183	70	36	303	0.676	0.602
CAMEROON																								
Douala	04°03'N	009°42'E	--	11:00:19.6	236	40	70	--			--			13:54:30.4	114	2	49	12:30:26.8	176	38	65	321	0.586	0.496
Garoua	09°18'N	013°24'E	--	11:22:58.8	227	62	75	--			--			13:51:43.2	127	24	48	12:40:20.8	177	58	63	305	0.409	0.298
Maroua	10°36'N	014°20'E	--	11:29:59.8	225	71	76	--			--			13:49:59.8	130	29	49	12:42:28.5	178	63	63	302	0.363	0.252
Yaoundé	03°52'N	011°31'E	770	11:05:21.4	238	50	70	--			--			13:58:11.4	115	4	46	12:35:22.3	177	44	63	317	0.588	0.498
CENTRAL AFRICAN REPUB																								
Bangui	04°22'N	018°35'E	387	11:28:31.7	238	87	69							14:07:58.9	120	17	38	12:53:00.6	180	62	54	305	0.538	0.440
CHAD																								
Moundou	08°34'N	016°05'E	--	11:29:21.2	230	78	73	--			--			13:57:09.7	127	25	45	12:46:46.2	179	64	60	303	0.421	0.310
Ndjamena	12°07'N	015°03'E	295	11:36:00.3	221	79	76	--			--			13:46:52.0	134	35	49	12:43:56.4	178	67	63	299	0.311	0.201
CISKEI																								
Potsdam (Mdants...	32°56'S	027°42'E	--	11:39:29.8	294	135	30							14:06:15.5	67	294	11	12:57:08.3	0	217	21	319	0.587	0.495
CONGO																								
Brazzaville	04°16'S	015°17'E	318	11:06:23.7	253	76	62	--			--			14:11:24.5	103	348	36	12:44:28.4	178	44	52	319	0.822	0.785
Pointe-Noire	04°48'S	011°51'E	50	10:54:51.8	253	62	61	--			--			14:05:30.4	100	341	40	12:34:43.9	177	34	56	326	0.851	0.822
DEMOCRATIC REPUBLIC OF CONGO																								
Kananga	05°54'S	022°25'E	--	11:28:27.4	257	104	58	--			--			14:21:38.6	105	354	27	13:01:40.6	181	57	43	310	0.820	0.783
Kinshasa	04°18'S	015°18'E	--	11:06:25.2	253	76	62	--			--			14:11:26.5	102	348	36	12:44:30.5	178	44	52	319	0.823	0.787
Kisangani	00°30'N	025°12'E	--	11:42:42.1	247	115	63	--			--			14:20:35.0	116	13	27	13:07:28.4	182	68	43	304	0.633	0.550
Kolwezi	10°43'S	025°28'E	--	11:35:07.8	264	115	51	--			--			14:25:29.5	99	346	21	13:07:10.0	182	57	37	310	0.930	0.922
Lubumbashi	11°40'S	027°28'E	--	11:40:38.7	265	119	49	--			--			14:27:16.3	99	347	18	13:07:16.3	183	57	34	309	0.937	0.931
Mbuji-Mayi	06°09'S	023°38'E	--	11:32:10.2	257	107	56	--			--			14:23:01.0	105	356	25	13:04:15.1	182	59	42	309	0.816	0.778
DJIBOUTI																								
Djibouti	11°36'N	043°09'E	7	13:05:16.9	202	111	33							13:48:30.6	167	80	23	13:27:15.6	185	96	28	291	0.050	0.014
EQUATORIAL GUINEA																								
Malabo	03°45'N	008°47'E	--	10:57:07.4	236	35	69							13:53:03.3	113	359	50	12:27:50.8	175	34	65	324	0.597	0.508
ETHIOPIA																								
Adis Abeba (Add...	09°02'N	038°42'E	2450	12:40:46.9	219	120	42	--			--			14:04:37.0	150	60	22	13:24:25.7	184	91	32	292	0.185	0.094
Dire Dawa	09°37'N	041°52'E	--	12:51:54.3	213	118	36	--			--			14:00:11.3	156	68	21	13:27:09.4	185	93	28	292	0.127	0.054
GABON																								
Libreville	00°23'N	009°27'E	35	10:53:28.0	243	44	66	--			--			13:57:57.0	107	351	46	12:29:00.6	175	32	62	327	0.700	0.633
Port-Gentil	00°43'S	008°47'E	--	10:49:57.9	244	43	64	--			--			13:57:23.8	105	347	46	12:26:47.6	175	28	62	330	0.734	0.676
GAMBIA																								
Banjul	13°28'N	016°39'W	--	10:51:01.7	189	290	56							12:08:00.6	139	262	73	11:28:38.8	164	272	64	63	0.101	0.038
GHANA																								
Accra	05°33'N	000°13'W	27	10:38:41.9	227	355	63	--			--			13:28:25.3	113	346	63	12:02:57.4	170	350	72	360	0.516	0.415
Kumasi	06°41'N	001°35'W	287	10:38:47.9	224	348	63	--			--			13:22:39.2	115	345	65	11:59:46.1	169	342	73	7	0.472	0.366
Tamale	09°25'N	000°50'W	--	10:47:52.0	219	343	67	--			--			13:19:41.9	121	356	67	12:03:15.3	170	348	76	2	0.391	0.279
Tema	05°38'N	000°01'E	--	10:39:22.6	227	356	63	--			--			13:28:58.1	113	346	62	12:03:39.3	170	352	72	359	0.515	0.414

Location Name	Latitude	Longitude	Elev. (m)	First Contact U.T. h m s	P °	V °	Alt	Second Contact U.T. h m s	P °	V °	Third Contact U.T. h m s	P °	V °	Fourth Contact U.T. h m s	P °	V °	Alt	Maximum Eclipse U.T. h m s	P °	V °	Alt	Azm	Eclip. Mag.	Eclip. Obs.
GUINEA																								
Conakry	09°31'N	013°43'W	7	10:32:38.2	205	311	53	—	—	—	—	—	—	12:34:25.7	124	283	75	11:31:19.7	164	285	65	53	0.260	0.155
Kankan	10°23'N	009°18'W	—	10:39:05.2	208	318	59	—	—	—	—	—	—	12:49:15.8	124	315	77	11:42:19.2	166	298	71	44	0.287	0.179
Labé	11°19'N	012°17'W	—	10:40:30.9	202	308	57	—	—	—	—	—	—	12:35:11.2	128	290	77	11:36:07.0	165	286	68	53	0.223	0.124
GUINEA-BISSAU																								
Bissau	11°51'N	015°35'W	—	10:41:58.4	196	299	54	—	—	—	—	—	—	12:19:39.5	132	268	74	11:29:21.7	164	277	65	60	0.164	0.079
IVORY COAST																								
Abidjan	05°19'N	004°02'W	20	10:30:59.5	224	345	59	—	—	—	—	—	—	13:17:08.9	112	333	67	11:52:13.9	168	328	71	18	0.496	0.392
Bouaké	07°41'N	005°02'W	364	10:35:38.4	218	336	60	—	—	—	—	—	—	13:10:02.8	117	335	70	11:51:06.0	167	321	73	24	0.414	0.303
Daloa	06°53'N	006°27'W	—	10:31:19.0	219	335	58	—	—	—	—	—	—	13:06:42.9	116	327	71	11:46:50.4	167	315	71	29	0.425	0.315
Korhogo	09°27'N	005°38'W	—	10:40:09.6	214	329	61	—	—	—	—	—	—	13:04:27.9	121	337	73	11:50:44.1	167	316	74	29	0.353	0.241
Yamoussoukro	06°49'N	005°17'W	—	10:32:50.6	220	338	59	—	—	—	—	—	—	13:10:44.8	115	332	70	11:49:51.6	167	320	72	25	0.438	0.329
KENYA																								
Kisumu	00°06'S	034°45'E	—	12:11:28.2	243	125	47	—	—	—	—	—	—	14:24:33.8	125	26	18	13:22:38.8	184	79	32	298	0.514	0.411
Mombasa	04°03'S	039°40'E	16	12:18:53.4	248	131	39	—	—	—	—	—	—	14:29:23.7	121	22	11	13:28:41.3	185	79	24	298	0.570	0.476
Nairobi	01°17'S	036°49'E	1820	12:15:21.5	244	128	44	—	—	—	—	—	—	14:26:20.2	124	26	15	13:25:23.1	184	79	29	298	0.524	0.423
Nakuru	00°17'S	036°04'E	—	12:14:44.3	243	126	45	—	—	—	—	—	—	14:24:58.6	125	28	17	13:24:19.1	184	79	30	298	0.503	0.399
LESOTHO																								
Maseru	29°28'S	027°30'E	—	11:38:38.5	290	132	33	—	—	—	—	—	—	14:11:45.7	71	302	12	13:00:05.9	1	220	23	318	0.658	0.581
LIBERIA																								
Monrovia	06°18'N	010°47'W	23	10:24:26.5	215	327	52	—	—	—	—	—	—	12:52:19.6	115	302	73	11:35:27.8	165	298	66	42	0.395	0.284
MALAWI																								
Blantyre	15°47'S	035°00'E	—	11:58:33.7	269	133	38	—	—	—	—	—	—	14:31:36.8	97	346	9	13:21:08.3	184	63	24	305	0.967	0.968
Lilongwe	13°59'S	033°44'E	—	11:56:25.2	267	131	41	—	—	—	—	—	—	14:31:22.5	99	349	11	13:20:06.3	184	64	26	305	0.932	0.925
Nsanje	16°55'S	035°12'E	—	11:58:29.5	271	134	38	—	—	—	—	—	—	14:31:19.0	96	343	9	13:20:54.4	184	62	23	305	0.996	0.998
Port Herald	16°54'S	035°16'E	—	11:58:39.1	271	134	37	—	—	—	—	—	—	14:31:21.5	96	343	8	13:21:00.1	184	62	23	305	0.995	0.997
MALI																								
Bamako	12°39'N	008°00'W	340	10:49:14.7	204	313	63	—	—	—	—	—	—	12:47:37.3	130	327	79	11:47:09.4	166	299	75	44	0.229	0.129
MAYOTTE																								
Dzaoudzi	12°47'S	045°17'E	—	12:21:42.4	259	138	29	—	—	—	—	—	—	14:34:08.1	110	6	1	13:32:31.9	185	74	14	299	0.754	0.699
NAMIBIA																								
Windhoek	22°34'S	017°06'E	1728	11:08:42.4	281	106	44	—	—	—	—	—	—	14:05:32.1	75	304	24	12:42:10.6	358	210	37	328	0.745	0.689
NIGER																								
Maradi	13°29'N	007°06'E	216	11:18:39.2	215	15	79	—	—	—	—	—	—	13:29:08.3	133	28	61	12:25:03.3	174	47	74	311	0.284	0.176
Niamey	13°31'N	002°07'E	—	11:07:39.3	212	343	75	—	—	—	—	—	—	13:17:10.0	131	18	68	12:12:39.5	172	17	79	336	0.272	0.166
Zinder	13°48'N	008°59'E	—	11:24:35.6	215	34	80	—	—	—	—	—	—	13:31:54.5	135	32	59	12:29:39.9	175	55	72	305	0.273	0.166
NIGERIA																								
Abeokuta	07°10'N	003°26'E	—	10:50:38.6	227	4	69	—	—	—	—	—	—	13:35:42.1	117	359	59	12:13:51.8	172	14	73	340	0.482	0.377
Ado-Ekiti	07°38'N	005°12'E	233	10:56:04.5	227	11	71	—	—	—	—	—	—	13:39:10.4	119	5	58	12:18:52.2	173	24	72	331	0.472	0.365
Enugu	06°27'N	007°27'E	—	10:59:02.4	231	24	71	—	—	—	—	—	—	13:46:18.4	118	5	54	12:24:43.4	174	33	69	325	0.512	0.410
Ibadan	07°17'N	003°30'E	—	10:51:05.6	226	9	70	—	—	—	—	—	—	13:35:39.9	118	360	59	12:14:05.3	173	14	73	339	0.478	0.373
Ilesha	07°38'N	004°45'E	—	10:54:58.1	227	6	70	—	—	—	—	—	—	13:38:06.4	119	4	58	12:17:38.4	173	22	72	333	0.471	0.364
Ilorin	08°30'N	004°32'E	—	10:56:42.8	225	6	71	—	—	—	—	—	—	13:35:55.7	121	359	59	12:17:19.0	173	22	73	333	0.442	0.333
Iwo	07°38'N	004°11'E	—	10:53:44.0	226	6	70	—	—	—	—	—	—	13:36:44.0	119	3	59	12:16:05.3	173	19	73	336	0.469	0.363
Kaduna	10°33'N	007°27'E	—	11:09:47.5	222	19	76	—	—	—	—	—	—	13:37:59.9	126	18	57	12:25:29.7	174	41	72	317	0.380	0.268
Kano	12°00'N	008°30'E	—	11:17:03.3	219	27	78	—	—	—	—	—	—	13:36:23.0	130	25	57	12:28:20.5	175	49	72	311	0.332	0.221
Lagos	06°27'N	003°24'E	3	10:48:48.6	228	6	68	—	—	—	—	—	—	13:36:49.5	116	357	59	12:13:30.4	172	13	72	341	0.505	0.402
Maiduguri	11°51'N	013°10'E	—	11:29:33.7	222	32	77	—	—	—	—	—	—	13:44:56.8	132	32	51	12:39:38.2	177	63	66	301	0.327	0.216
Mushin	06°32'N	003°22'E	—	10:48:55.9	228	5	68	—	—	—	—	—	—	13:36:36.3	116	357	59	12:13:26.6	172	13	72	341	0.502	0.399
Ogbomosho	08°08'N	004°15'E	—	10:55:53.5	225	5	71	—	—	—	—	—	—	13:35:57.5	120	4	59	12:16:26.0	172	20	73	335	0.453	0.345
Onitsha	06°09'N	006°47'E	—	10:56:34.9	231	21	70	—	—	—	—	—	—	13:45:20.4	117	3	54	12:22:49.3	174	30	69	327	0.521	0.420
Oshogbo	07°47'N	004°34'E	—	10:54:54.4	226	8	70	—	—	—	—	—	—	13:37:23.3	119	4	59	12:17:11.1	173	21	72	333	0.465	0.358
Port Harcourt	04°43'N	007°05'E	—	10:54:18.0	234	25	69	—	—	—	—	—	—	13:48:09.0	114	359	53	12:23:18.1	174	29	68	328	0.566	0.472
Zaria	11°07'N	007°44'E	—	11:12:14.9	221	21	77	—	—	—	—	—	—	13:37:10.8	128	21	58	12:26:18.3	174	44	72	315	0.361	0.250

TABLE 10 - CONTINUED
LOCAL CIRCUMSTANCES FOR AFRICA
TOTAL SOLAR ECLIPSE OF 2001 JUNE 21

Location Name	Latitude	Longitude	Elev. (m)	First Contact U.T. h m s	P°	V°	Alt	Second Contact U.T.	P°	V°	Third Contact U.T.	P°	V°	Fourth Contact U.T. h m s	P°	V°	Alt	Maximum Eclipse U.T. h m s	P°	V°	Alt	Azm	Eclip. Mag.	Eclip. Obs.
RWANDA																								
Kigali	01°57'S	030°04'E	–	11:55:53.7	249	120	52	–			–			14:25:06.5	117	15	21	13:16:08.4	183	72	36	301	0.625	0.541
SENEGAL																								
Dakar	14°40'N	017°26'W	40	11:00:22.9	182	282	58	–			–			11:56:51.4	146	259	70	11:28:12.2	164	269	64	66	0.054	0.015
Kaolack	14°09'N	016°04'W	–	10:55:03.1	187	288	58	–			–			12:07:21.5	141	261	73	11:30:28.7	164	272	66	64	0.088	0.031
Saint-Louis	16°02'N	016°30'W	–	11:02:07.0	176	276	62	–			–			11:50:49.6	152	260	71	11:31:21.4	164	268	66	66	0.025	0.005
Thiès	14°48'N	016°56'W	–	11:00:44.9	182	283	58	–			–			11:58:46.7	145	260	71	11:29:19.7	164	270	65	66	0.057	0.016
Ziguinchor	12°35'N	016°16'W	–	10:45:47.0	192	294	55	–			–			12:13:47.5	135	264	73	11:28:36.5	164	275	64	62	0.133	0.058
SIERRA LEONE																								
Freetown	08°30'N	013°15'W	28	10:29:15.7	208	315	52	–			–			12:38:41.7	121	286	75	11:31:29.7	164	288	65	51	0.297	0.188
SOMALIA																								
Muqdisho (Mogad…	02°04'N	045°22'E	12	12:41:03.8	231	127	33	–			–			14:19:19.4	139	47	10	13:32:41.9	185	88	21	294	0.311	0.200
SOUTH AFRICA																								
Alexandria	33°39'S	026°24'E	–	11:36:47.7	295	134	30	–			–			14:03:11.0	65	290	12	12:54:03.4	0	215	22	321	0.563	0.467
Benoni	26°19'S	028°27'E	–	11:40:40.1	286	131	35	–			–			14:17:26.8	77	311	12	13:04:30.6	1	225	24	315	0.737	0.678
Bloemfontein	29°12'S	026°07'E	–	11:35:21.2	290	130	34	–			–			14:10:21.3	71	301	13	12:57:45.1	1	219	25	319	0.653	0.575
Boksburg	26°12'S	028°14'E	–	11:40:08.9	285	131	35	–			–			14:17:21.0	77	311	12	13:04:15.2	1	225	25	316	0.738	0.679
Cape Town (Kaap…	33°55'S	018°22'E	17	11:17:46.2	295	122	32	–			–			13:49:28.9	60	278	18	12:36:51.8	358	203	27	332	0.513	0.410
Carletonville	26°23'S	027°22'E	–	11:38:03.5	286	130	36	–			–			14:16:04.3	76	309	12	13:02:29.8	1	223	25	315	0.727	0.665
Daveyton Locati…	26°09'S	028°25'E	–	11:40:35.2	285	131	36	–			–			14:17:37.6	77	311	12	13:04:35.2	1	225	25	315	0.741	0.683
Durban	29°55'S	030°56'E	5	11:46:12.3	290	137	31	–			–			14:15:07.8	73	305	9	13:05:31.5	0	224	20	314	0.677	0.604
East London (Oo…	33°00'S	027°55'E	–	11:39:58.1	294	136	30	–			–			14:06:25.7	67	294	10	12:57:27.3	1	217	21	319	0.587	0.495
Evaton	26°31'S	027°54'E	–	11:39:21.5	286	131	36	–			–			14:16:32.2	76	310	12	13:03:21.9	1	224	25	316	0.728	0.666
Germiston	26°13'S	028°10'E	–	11:39:59.3	286	131	36	–			–			14:17:15.0	77	311	12	13:04:05.2	1	225	25	316	0.737	0.678
Johannesburg	26°15'S	028°00'E	–	11:39:35.3	286	131	36	–			–			14:17:00.6	76	310	12	13:03:45.6	1	225	25	316	0.735	0.675
Kempton Park	26°06'S	028°14'E	–	11:40:08.7	286	131	36	–			–			14:17:28.8	77	311	12	13:04:18.0	1	225	25	316	0.741	0.682
Kimberley	28°43'S	024°46'E	1197	11:31:58.9	289	128	35	–			–			14:09:14.5	71	300	15	12:55:31.9	0	218	26	321	0.653	0.575
Klerksdorp	26°58'S	026°39'E	–	11:36:21.0	287	129	36	–			–			14:14:22.4	74	307	13	13:00:40.8	1	222	26	318	0.707	0.641
Mamelodi	25°45'S	028°18'E	–	11:40:18.0	285	131	36	–			–			14:18:00.6	77	312	12	13:04:41.9	2	225	25	315	0.749	0.693
Mdantsana (Pots…	32°56'S	027°42'E	–	11:39:29.8	294	135	30	–			–			14:06:15.5	67	294	11	12:57:08.3	0	217	21	319	0.587	0.495
Natalspruit	26°19'S	028°09'E	–	11:39:57.1	286	131	36	–			–			14:17:05.9	76	311	12	13:03:58.6	1	225	25	315	0.735	0.675
Pietermaritzburg	29°37'S	030°16'E	–	11:44:48.1	289	136	31	–			–			14:14:50.9	73	305	9	13:04:44.0	1	224	21	319	0.678	0.605
Port Elizabeth	33°58'S	025°40'E	58	11:35:15.0	295	134	30	–			–			14:03:15.6	64	288	12	12:52:21.8	360	214	22	322	0.552	0.454
Pretoria	25°45'S	028°10'E	1369	11:39:59.9	285	131	36	–			–			14:17:52.5	77	312	12	13:04:28.9	1	225	25	315	0.748	0.691
Soweto	26°14'S	027°54'E	–	11:39:20.8	286	131	36	–			–			14:16:54.8	76	310	12	13:03:35.6	1	224	25	316	0.735	0.675
Vereeniging	26°38'S	027°57'E	–	11:39:29.0	286	131	35	–			–			14:16:26.3	76	309	12	13:03:21.6	1	224	25	316	0.726	0.664
SUDAN																								
Khartoum	15°36'N	032°32'E	390	12:48:30.6	201	109	48	–			–			13:38:03.4	165	77	36	13:13:44.9	183	93	42	288	0.054	0.015
Al-Khartum Bahri	15°38'N	032°33'E	–	12:48:48.0	201	109	48	–			–			13:37:47.3	165	77	36	13:13:44.9	183	93	42	288	0.053	0.015
Umm Durman	15°38'N	032°30'E	–	12:48:35.6	201	109	48	–			–			13:37:51.1	165	77	37	13:13:40.8	183	93	42	288	0.053	0.015
SWAZILAND																								
Mbabane	26°18'S	031°06'E	–	11:46:46.2	285	135	34	–			–			14:20:18.9	78	314	9	13:08:58.1	2	228	22	313	0.762	0.709
TANZANIA																								
Dar es Salaam	06°48'S	039°17'E	14	12:15:01.2	253	132	39	–			–			14:31:33.1	116	14	9	13:28:16.0	185	76	23	299	0.657	0.579
Mbeya	08°54'S	033°27'E	–	11:58:52.7	259	128	45	–			–			14:30:57.9	108	14	9	13:20:58.3	184	68	19	303	0.791	0.745
Mwanza	02°31'S	032°54'E	–	12:03:20.9	248	124	48	–			–			14:26:49.3	119	18	18	13:20:26.3	184	74	33	300	0.609	0.522
Tabora	05°01'S	032°48'E	–	12:00:26.4	253	125	47	–			–			14:28:49.3	114	11	17	13:20:22.4	184	72	32	301	0.685	0.613
Tanga	05°04'S	039°06'E	–	12:16:24.4	250	131	39	–			–			14:30:16.5	119	11	11	13:28:07.6	185	77	24	299	0.608	0.520
Zanzibar	06°10'S	039°11'E	–	12:15:25.9	252	132	39	–			–			14:31:07.1	117	16	10	13:28:11.6	185	76	24	299	0.640	0.558
TOGO																								
Lomé	06°08'N	001°13'E	22	10:43:06.7	227	358	65	–			–			13:31:34.1	115	351	61	12:07:15.8	171	359	73	352	0.506	0.403
TRANSKEI																								
Umtata	31°35'S	028°47'E	–	11:41:41.2	292	136	30	–			–			14:09:58.3	69	298	10	13:00:22.4	1	220	21	317	0.623	0.538
VENDA																								
Thohoyandou	23°00'S	030°29'E	–	11:45:41.7	281	132	37	–			–			14:23:28.0	83	322	11	13:10:28.3	2	231	24	312	0.836	0.803

TABLE 11
LOCAL CIRCUMSTANCES FOR ANGOLA
TOTAL SOLAR ECLIPSE OF 2001 JUNE 21

Location Name	Latitude	Longitude	Elev. m	First Contact U.T. h m s	P °	V °	Alt °	Second Contact U.T. h m s	P °	V °	Third Contact U.T. h m s	P °	V °	Fourth Contact U.T. h m s	P °	V °	Alt °	Maximum Eclipse U.T. h m s	P °	V °	Alt °	Azm °	Eclip. Mag.	Eclip. Obs.	Umbral Depth	Umbral Durat.
ANGOLA																										
Alto-Uama	12°14'S	015°33'E	—	11:02:54.8	266	87	54	12:42:12.6	156	14	12:43:52.4	200	58	14:11:37.5	90	328	31	12:43:02.7	358	216	46	325	1.002	1.000	0.072	01m40s
Amboíva	11°32'S	014°44'E	—	11:00:25.5	264	83	55	12:38:49.5	104	321	12:43:10.7	251	109	14:10:27.1	90	328	32	12:41:00.4	358	215	48	326	1.017	1.000	0.716	04m21s
Andulo	11°30'S	016°45'E	—	11:07:01.7	265	90	55	12:44:23.9	79	300	12:48:45.3	278	140	14:13:56.2	92	332	30	12:46:34.9	179	40	46	322	1.020	1.000	0.834	04m21s
Bailundo	12°12'S	015°52'E	—	11:03:57.3	266	88	54	12:42:40.6	144	3	12:45:10.9	212	72	14:12:11.4	90	328	31	12:43:56.0	358	217	46	324	1.004	1.000	0.173	02m30s
Benguela	12°35'S	013°25'E	—	10:55:54.1	265	80	54	—			—			14:07:32.5	88	323	33	12:36:49.1	357	211	48	329	0.984	0.987		
Bimbe	11°49'S	015°49'E	—	11:03:53.3	265	87	55	12:41:52.8	112	330	12:45:58.6	245	104	14:12:15.3	91	329	31	12:43:56.0	358	217	46	324	1.014	1.000	0.600	04m06s
Botera	11°37'S	014°17'E	—	10:58:56.1	264	82	55	12:37:39.5	115	330	12:41:43.4	241	97	14:09:36.0	90	327	33	12:39:41.7	358	214	48	326	1.013	1.000	0.545	04m04s
Branco	12°30'S	020°32'E	—	11:19:01.8	267	103	53	12:53:46.2	88	314	12:57:54.8	273	140	14:19:18.5	92	335	25	12:55:50.9	180	47	41	317	1.022	1.000	0.952	04m09s
Calala	12°59'S	023°30'E	—	11:28:13.6	268	111	51	13:00:43.6	60	290	13:04:06.4	302	173	14:22:55.7	94	338	22	13:02:25.3	181	52	38	314	1.011	1.000	0.486	03m23s
Calucinga	11°18'S	016°12'E	—	11:05:17.3	264	88	55	12:43:00.8	73	292	12:47:18.1	284	145	14:13:05.3	92	331	31	12:45:09.8	179	39	46	323	1.018	1.000	0.728	04m17s
Camacupa	12°03'S	017°30'E	—	11:09:19.7	266	93	54	12:46:17.5	106	327	12:50:28.0	252	115	14:14:57.7	91	331	29	12:48:23.1	359	221	44	321	1.017	1.000	0.713	04m10s
Cangumbe	12°00'S	019°17'E	—	11:15:08.6	266	99	54	12:50:54.5	75	300	12:55:00.9	284	150	14:17:42.2	92	334	27	12:52:58.0	180	45	43	318	1.018	1.000	0.749	04m06s
Capage	13°21'S	021°05'E	—	11:20:34.0	268	105	52	12:55:32.6	141	7	12:58:09.0	220	87	14:19:44.4	91	333	24	12:56:51.0	0	227	40	317	1.005	1.000	0.229	02m36s
Cariango	10°37'S	015°20'E	—	11:02:40.4	263	84	56	12:41:54.1	27	246	12:44:06.3	329	189	14:11:48.4	92	332	32	12:43:00.4	178	37	48	324	1.003	1.000	0.126	02m12s
Cassamba	13°06'S	020°18'E	—	11:18:07.8	268	103	52	12:53:38.6	137	2	12:56:29.9	224	89	14:18:45.6	91	333	25	12:55:04.6	0	226	41	318	1.006	1.000	0.274	02m51s
Cassongue	11°51'S	015°03'E	—	11:01:22.5	265	85	55	12:39:57.3	124	341	12:43:36.6	232	90	14:10:53.9	90	328	32	12:41:47.2	358	216	47	325	1.010	1.000	0.413	03m39s
Caxopa	11°52'S	020°52'E	—	11:20:17.1	266	103	53	12:55:56.2	27	254	12:57:45.3	335	202	14:19:57.4	94	337	25	12:56:51.0	180	48	41	316	1.002	1.000	0.102	01m49s
Cela	11°25'S	015°07'E	—	11:01:42.6	264	84	55	12:39:52.2	93	310	12:44:22.6	263	122	14:11:10.7	91	329	32	12:42:07.7	358	216	47	325	1.022	1.000	0.919	04m30s
Chenele	12°54'S	023°54'E	—	11:29:29.0	268	112	51	13:02:02.7	41	272	13:04:33.4	322	193	14:23:25.0	94	339	22	13:03:18.3	181	52	37	313	1.005	1.000	0.233	02m31s
Chepaúa	12°58'S	022°43'E	—	11:25:48.1	268	109	51	12:58:46.5	77	306	13:02:38.8	285	155	14:21:59.5	93	337	23	13:00:43.0	181	50	38	315	1.018	1.000	0.758	03m52s
Chibango	13°38'S	021°56'E	—	11:23:10.8	269	108	51	12:57:33.6	146	13	12:59:51.7	216	83	14:20:45.0	92	334	23	12:58:42.9	1	228	39	316	1.004	1.000	0.179	02m18s
Chiengo	13°20'S	021°55'E	—	11:23:12.0	268	108	51	12:57:00.2	119	346	13:00:33.6	243	111	14:20:51.0	92	335	24	12:58:47.3	1	228	39	316	1.012	1.000	0.527	03m33s
Chila	12°04'S	014°29'E	—	10:59:28.4	265	83	54	12:39:16.7	156	12	12:40:55.2	199	55	14:09:47.0	89	326	32	12:40:06.1	358	214	47	326	1.002	1.000	0.068	01m39s
Coemba	12°08'S	018°05'E	—	11:11:12.5	266	95	54	12:47:46.4	103	325	12:51:58.2	256	119	14:15:51.0	91	332	28	12:49:52.7	359	222	44	320	1.018	1.000	0.765	04m12s
Conda	11°06'S	014°20'E	—	10:59:14.8	263	81	55	12:37:45.2	80	296	12:42:16.9	276	133	14:09:52.7	91	329	33	12:40:01.4	178	34	48	326	1.021	1.000	0.860	04m32s
Condé	10°50'S	014°37'E	—	11:00:15.4	263	82	56	12:38:55.8	58	275	12:42:54.2	297	156	14:10:28.7	91	330	33	12:40:55.3	178	35	48	325	1.012	1.000	0.508	03m58s
Copolo	10°22'S	014°07'E	—	10:58:47.3	262	79	56	12:38:50.8	18	235	12:40:27.7	337	194	14:09:42.3	92	330	34	12:39:39.4	178	35	49	326	1.002	1.000	0.063	01m37s
Curunga	12°51'S	021°12'E	—	11:21:03.6	268	105	52	12:55:16.2	99	326	12:59:19.4	262	130	14:20:05.4	92	335	25	12:57:18.2	1	228	40	316	1.020	1.000	0.854	04m03s
Dala Cachibo	10°28'S	014°39'E	—	11:00:29.5	263	81	56	12:40:13.0	22	239	12:42:03.5	334	192	14:10:38.6	92	331	33	12:41:08.4	178	36	48	325	1.002	1.000	0.085	01m50s
Ebo	11°02'S	014°41'E	—	11:00:24.6	263	82	56	12:38:50.9	72	289	12:43:13.5	284	142	14:10:32.1	91	329	33	12:41:02.5	178	35	48	325	1.018	1.000	0.724	04m23s
Ferreira	12°53'S	022°48'E	—	11:26:05.2	268	110	51	12:59:04.4	69	298	13:02:46.5	293	163	14:22:07.3	93	337	23	13:00:55.8	181	50	38	314	1.015	1.000	0.629	03m42s
Gabela	10°48'S	014°20'E	—	10:59:20.6	263	81	56	12:38:06.9	59	275	12:42:07.9	296	154	14:09:58.5	91	329	33	12:40:07.7	178	35	49	326	1.013	1.000	0.519	04m01s
Gando	12°30'S	017°25'E	—	11:08:56.8	266	93	54	12:46:46.4	145	6	12:49:13.9	213	75	14:14:39.2	90	330	29	12:48:00.4	359	220	44	322	1.004	1.000	0.175	02m27s
Gumba	11°40'S	016°34'E	—	11:06:22.9	265	90	55	12:43:49.0	93	313	12:48:14.0	265	126	14:13:34.7	91	331	30	12:46:01.9	359	219	46	322	1.022	1.000	0.933	04m25s
Gungo	11°48'S	014°08'E	—	10:58:24.0	264	82	55	12:37:32.1	131	346	12:40:51.2	224	80	14:09:15.0	90	327	33	12:39:11.9	358	213	48	327	1.008	1.000	0.313	03m19s
Huambo	12°44'S	015°47'E	—	11:03:34.1	266	88	54	—			—			14:11:48.7	89	327	31	12:43:30.3	358	216	46	325	0.989	0.992		
Kuito	12°22'S	016°56'E	—	11:07:24.0	266	92	54	12:45:24.6	141	1	12:48:07.2	217	78	14:13:55.2	90	330	29	12:46:46.2	359	219	45	322	1.005	1.000	0.214	02m43s
Limbueta	12°30'S	018°42'E	—	11:13:07.1	267	97	54	12:49:28.0	119	342	12:53:10.8	240	104	14:16:39.6	91	332	27	12:51:19.7	360	223	43	320	1.012	1.000	0.506	03m43s
Lobito	12°20'S	013°34'E	—	10:56:26.2	265	80	54	—			—			14:07:56.9	88	324	33	12:37:21.3	357	211	48	328	0.991	0.994		
Luanda	08°48'S	013°14'E	59	10:56:35.4	259	74	58	—			—			14:08:23.0	94	333	36	12:37:37.2	177	34	51	326	0.961	0.961		
Lucusse	12°32'S	020°48'E	—	11:19:52.4	267	104	53	12:54:24.5	85	311	12:58:31.2	276	143	14:19:39.7	93	335	25	12:56:28.3	180	47	41	317	1.021	1.000	0.905	04m07s
Lumbala	12°39'S	022°34'E	—	11:25:25.6	267	109	52	12:58:49.8	57	286	13:02:08.5	305	175	14:21:54.7	94	337	23	13:00:29.5	181	50	39	315	1.010	1.000	0.437	03m19s
Lunge	12°12'S	016°05'E	—	11:04:39.8	266	89	54	12:43:08.5	140	359	12:45:54.5	217	77	14:12:33.7	90	329	30	12:44:31.7	358	218	46	324	1.005	1.000	0.218	02m46s
Luso	11°47'S	019°52'E	—	11:17:05.7	266	100	54	12:52:55.5	47	273	12:56:01.1	313	179	14:18:36.4	93	336	26	12:54:28.6	180	46	42	318	1.008	1.000	0.322	03m06s
Moçâmedes	15°10'S	012°09'E	—	10:51:35.8	269	80	51	—			—			14:03:24.1	83	315	33	12:31:57.4	356	205	47	333	0.912	0.899		
Muangai	12°32'S	019°51'E	—	11:16:49.6	267	101	53	12:52:07.4	102	327	12:56:13.8	258	124	14:18:20.3	92	334	26	12:54:11.0	0	225	42	318	1.019	1.000	0.798	04m06s
Munhango	12°12'S	018°42'E	—	11:13:11.7	266	97	54	12:49:18.7	98	321	12:53:32.5	261	126	14:16:46.2	92	333	28	12:51:26.0	360	223	43	319	1.020	1.000	0.853	04m14s
Ngunza	11°13'S	013°50'E	—	10:57:35.1	263	80	55	12:36:14.8	92	306	12:40:50.0	263	119	14:08:54.8	90	328	34	12:38:32.7	357	213	49	327	1.022	1.000	0.927	04m35s
Nhareia	11°25'S	017°03'E	—	11:08:02.1	265	91	55	12:45:19.8	70	291	12:49:28.9	288	151	14:14:27.2	92	332	30	12:47:24.7	179	41	45	321	1.016	1.000	0.667	04m09s
Nova Sintra	12°09'S	017°16'E	—	11:08:32.5	266	93	54	12:45:47.2	116	337	12:49:40.2	242	104	14:14:33.0	91	331	29	12:47:44.1	359	220	45	322	1.013	1.000	0.541	03m53s
Porto Amboim	10°44'S	013°44'E	—	10:57:24.7	263	79	56	12:36:22.3	60	275	12:40:28.4	295	151	14:08:53.3	91	329	34	12:38:25.7	177	33	49	327	1.013	1.000	0.541	04m06s
Quibala	10°46'S	014°59'E	—	11:01:28.6	263	83	56	12:40:13.1	49	267	12:43:43.8	307	166	14:11:09.1	92	331	32	12:41:58.7	178	36	48	325	1.009	1.000	0.365	03m31s
Quilenda	10°33'S	014°22'E	—	10:59:32.2	263	81	56	12:38:50.6	37	254	12:41:46.0	318	176	14:10:06.6	92	330	33	12:40:18.5	178	35	49	326	1.006	1.000	0.229	02m55s
Quirima	10°48'S	018°09'E	—	11:11:50.3	264	94	55	—			—			14:16:20.9	94	336	29	12:50:29.2	179	44	45	319	0.992	0.995		
Quirimbo	10°36'S	014°12'E	—	10:58:58.6	263	80	56	12:38:08.4	44	261	12:41:29.1	311	168	14:09:47.5	91	330	33	12:39:49.0	178	35	49	326	1.008	1.000	0.315	03m21s
Sacaola	12°57'S	022°25'E	—	11:24:52.2	268	109	51	12:58:03.9	82	310	13:02:01.8	280	149	14:21:37.6	93	336	23	13:00:03.3	181	50	39	315	1.020	1.000	0.846	03m58s
Sanga	11°07'S	015°22'E	—	11:02:37.0	264	85	55	12:40:46.8	70	288	12:45:04.3	286	146	14:11:43.1	91	330	32	12:42:55.9	178	37	47	324	1.017	1.000	0.691	04m17s
Sapu	12°29'S	019°26'E	—	11:15:29.8	267	100	53	12:51:08.1	105	330	12:55:12.2	254	120	14:17:45.3	92	333	27	12:53:10.5	360	225	42	319	1.017	1.000	0.733	04m04s
Saricumbe	12°12'S	019°46'E	—	11:16:38.9	266	100	53	12:52:00.7	81	306	12:56:09.8	279	145	14:18:20.1	92	334	26	12:54:05.6	180	46	42	318	1.020	1.000	0.839	04m09s
Uiche	12°03'S	021°02'E	—	11:20:45.6	266	104	53	12:55:47.2	43	270	12:58:34.1	318	186	14:20:07.7	94	337	25	12:57:11.0	181	48	41	316	1.006	1.000	0.262	02m47s
Umpulo	12°38'S	017°42'E	—	11:09:50.3	267	95	54	12:47:47.9	155	16	12:49:36.0	204	66	14:15:03.2	90	330	28	12:48:42.1	359	221	44	321	1.002	1.000	0.091	01m48s
Urimba	10°56'S	016°32'E	—	11:06:29.8	264	89	56	12:44:45.5	39	260	12:47:36.6	319	181	14:13:45.1	92	333	31	12:46:11.3	179	40	46	322	1.006	1.000	0.233	02m51s
Vila Nova do Se...	11°24'S	014°15'E	—	10:58:53.2	264	81	55	12:37:26.4	100	316	12:41:54.3	255	112	14:09:37.2	90	328	33	12:39:40.7	358	214	48	326	1.019	1.000	0.785	04m28s
Vouga	12°11'S	016°47'E	—	11:06:57.3	266	91	54	12:44:41.1	126	346	12:48:10.3	231	92	14:13:44.8	91	330	30	12:46:26.0	359	219	45	322	1.009	1.000	0.390	03m29s

TABLE 12
LOCAL CIRCUMSTANCES FOR MADAGASCAR
TOTAL SOLAR ECLIPSE OF 2001 JUNE 21

Location Name	Latitude	Longitude	Elev. (m)	First Contact U.T. (h m s)	P (°)	V (°)	Alt (°)	Second Contact U.T. (h m s)	P (°)	V (°)	Third Contact U.T. (h m s)	P (°)	V (°)	Fourth Contact U.T. (h m s)	P (°)	V (°)	Alt (°)	Maximum Eclipse U.T. (h m s)	P (°)	V (°)	Alt (°)	Azm (°)	Eclip. Mag.	Eclip. Obs.	Umbral Depth	Umbral Durat.
MADAGASCAR																										
Ambahikily	21°36'S	043°41'E	—	12:12:38.6	273	142	26	13:26:00.9	98	337	13:28:39.9	270	150					13:27:20.7	4	244	12	302	1.018	1.000	0.936	02m39s
Ambivy	21°31'S	044°02'E	—	12:13:15.7	273	142	26	13:26:24.4	79	319	13:28:57.0	290	170					13:27:41.0	184	64	12	301	1.014	1.000	0.732	02m33s
Amparihy	23°57'S	047°20'E	—	12:16:41.0	275	145	21											13:28:30.0	4	245	8	300	0.999	1.000		
Ampasibe	22°56'S	046°58'E	—	12:16:49.8	273	144	22	13:27:42.4	86	326	13:30:11.0	282	163					13:28:56.9	184	65	9	300	1.016	1.000	0.861	02m29s
Ampoza	22°20'S	044°44'E	—	12:13:52.9	274	143	25	13:26:32.6	119	359	13:28:53.7	249	129					13:27:43.4	4	244	11	301	1.011	1.000	0.573	02m21s
Analavoka	22°33'S	046°30'E	—	12:16:30.1	273	144	22	13:27:42.7	72	312	13:30:02.6	297	177					13:28:52.6	184	65	9	300	1.012	1.000	0.618	02m20s
Andranopasy	21°17'S	043°44'E	—	12:12:54.8	273	142	26	13:26:21.7	70	310	13:28:47.5	298	178					13:27:34.8	184	64	12	302	1.012	1.000	0.597	02m26s
Andriandampy	22°45'S	045°41'E	—	12:15:04.6	274	143	24	13:27:00.1	119	358	13:29:19.6	250	130					13:28:10.1	4	244	10	301	1.011	1.000	0.585	02m20s
Ankaramena	21°57'S	046°39'E	—	12:17:01.2	274	143	23											13:29:22.0	184	65	9	300	0.994	0.996		
Ankarimbelo	22°08'S	047°20'E	—	12:17:53.1	272	144	22											13:29:42.6	184	66	9	300	0.990	0.993		
Ankazoabo	22°18'S	044°31'E	—	12:13:33.8	274	143	25	13:26:27.1	125	5	13:28:41.4	243	123					13:27:34.5	4	244	11	301	1.009	1.000	0.485	02m14s
Antambohobe	22°20'S	046°47'E	—	12:16:57.7	273	144	23	13:28:35.5	34	275	13:29:50.0	335	216					13:29:12.9	184	65	9	300	1.002	1.000	0.131	01m14s
Antananarivo	18°55'S	047°31'E	—	12:20:26.3	267	142	24											13:31:39.3	185	69	10	299	0.900	0.882		
Antanimieva	22°12'S	043°44'E	—	12:12:22.6	274	142	26	13:26:19.4	153	32	13:27:41.0	215	94					13:27:00.4	4	243	12	302	1.003	1.000	0.142	01m22s
Antevamena	21°02'S	044°08'E	—	12:13:43.0	272	142	26											13:28:03.4	184	65	12	301	1.000	1.000		
Antsirabe	19°51'S	047°02'E	—	12:19:01.9	269	142	24											13:30:51.1	184	68	10	299	0.932	0.923		
Antsiranana	12°16'S	049°17'E	—	12:28:53.2	256	139	25											13:35:12.3	185	77	11	297	0.687	0.615		
Basibasy	22°10'S	043°40'E	—	12:12:17.2	274	142	26	13:26:16.6	152	32	13:27:39.7	215	95					13:26:58.3	4	243	12	302	1.003	1.000	0.148	01m23s
Befandriana	22°06'S	043°54'E	—	12:12:42.2	274	142	26	13:26:10.0	132	12	13:28:14.5	236	115					13:27:12.0	4	243	12	302	1.007	1.000	0.382	02m04s
Befotaka	23°49'S	046°59'E	—	12:16:17.0	275	144	22											13:28:21.6	4	244	8	300	0.998	0.999		
Befotaka	21°29'S	044°44'E	—	12:14:24.1	273	143	25	13:27:23.8	45	286	13:29:06.8	323	204					13:28:15.1	184	65	11	301	1.005	1.000	0.247	01m43s
Bekisopa	21°40'S	045°54'E	—	12:16:05.3	272	142	24											13:29:01.0	184	65	10	300	0.995	0.997		
Belo-Sur-Mer	20°44'S	044°00'E	—	12:13:41.0	271	142	26											13:28:07.6	184	65	12	301	0.994	0.996		
Bemarivo	21°45'S	044°45'E	—	12:14:15.8	273	143	25	13:26:54.3	71	311	13:29:18.0	297	178					13:28:06.4	184	64	11	301	1.012	1.000	0.607	02m24s
Bemavo	21°41'S	045°10'E	—	12:15:21.5	273	143	25	13:28:29.8	13	253	13:28:51.8	356	237					13:28:40.9	184	65	11	301	1.005	1.000	0.010	00m22s
Beroroha	22°27'S	045°44'E	—	12:15:24.3	273	143	25											13:28:28.0	184	65	11	301	0.996	0.998		
Betioky	22°27'S	043°44'E	—	12:12:13.9	275	143	26	13:27:36.3	46	286	13:29:19.3	323	203					13:26:50.6	4	243	12	302	1.002	1.000	0.252	01m43s
Betroka	23°16'S	046°06'E	—	12:13:57.5	273	144	23	13:27:30.8	155	35	13:28:43.8	213	92					13:28:29.5	184	64	12	301	1.009	1.000	0.123	01m13s
Betsioky	21°31'S	044°28'E	—	12:13:57.5	273	144	23	13:26:55.3	62	302	13:29:07.8	307	187					13:28:01.7	184	64	12	301	1.007	1.000	0.461	02m12s
Etrotroka	22°53'S	047°36'E	—	12:17:45.3	273	144	22	13:28:25.3	56	297	13:30:22.1	312	193					13:29:23.9	184	65	8	300	1.003	1.000	0.383	01m57s
Farafangana	22°49'S	047°50'E	—	12:18:07.4	273	144	22	13:28:58.0	35	275	13:30:12.6	334	215					13:29:35.5	184	65	8	300	1.003	1.000	0.137	01m15s
Fenoarivo	21°43'S	046°24'E	—	12:16:48.3	272	143	24											13:29:48.3	184	65	10	300	0.991	0.993		
Fianarantsoa	21°26'S	047°05'E	—	12:18:00.0	271	144	23											13:29:58.6	184	66	9	300	0.974	0.976		
Iakora	23°06'S	046°40'E	—	12:17:36.4	274	144	22	13:27:25.6	111	351	13:29:49.9	257	138					13:28:38.0	4	244	9	300	1.014	1.000	0.713	02m24s
Ihorombe	23°00'S	047°33'E	—	12:17:58.0	273	143	23	13:28:09.7	69	310	13:30:24.6	299	180					13:29:17.4	184	65	8	300	1.011	1.000	0.581	02m15s
Ihosy	22°24'S	046°08'E	—	12:15:58.0	273	143	23	13:27:31.9	73	313	13:29:54.1	295	176					13:27:52.4	4	243	9	301	1.006	1.000	0.641	02m22s
Itandrano	22°12'S	046°10'E	—	12:16:02.7	273	143	24	13:27:46.8	56	297	13:29:39.8	312	193					13:28:43.3	184	65	10	300	1.012	1.000	0.317	01m53s
Ivahona	23°27'S	046°10'E	—	12:15:21.1	275	144	23											13:28:02.8	4	244	9	301	0.998	0.999		
Ivohibe	22°29'S	046°54'E	—	12:17:57.5	273	144	23	13:28:17.5	49	290	13:30:03.3	320	200					13:29:10.4	184	64	9	301	1.000	1.000	0.289	01m46s
Jangany	23°14'S	045°27'E	—	12:14:25.8	275	144	24											13:27:40.3	4	243	10	301	0.996	0.998		
Karianga	22°22'S	047°26'E	—	12:17:52.1	272	144	22											13:29:37.6	184	65	9	300	0.995	0.997		
Lambomakondro	22°41'S	044°44'E	—	12:13:40.4	274	143	24	13:26:57.5	160	39	13:28:01.5	208	88					13:27:29.6	4	244	11	301	1.002	1.000	0.088	01m04s
Lavaraty	23°16'S	046°24'E	—	12:14:48.9	274	143	24	13:27:33.1	113	353	13:29:54.9	255	136					13:28:44.4	4	244	9	301	1.013	1.000	0.678	02m22s
Mahabo	23°25'S	044°17'E	—	12:15:10.1	275	143	23											13:27:52.4	4	243	9	301	0.992	0.995		
Mahajanga	15°43'S	046°19'E	—	12:21:01.6	263	140	27											13:32:17.4	185	72	12	299	0.825	0.788		
Mahasoa	22°12'S	046°06'E	—	12:14:57.8	272	142	24	13:27:49.4	56	297	13:29:49.3	312	193	14:32:01.5	94	340	0	13:28:39.5	184	65	11	301	1.007	1.000	0.381	02m00s
Mandabe	21°03'S	044°55'E	—	12:14:57.8	272	142	25											13:28:39.7	184	65	11	301	0.991	0.994		
Manera	22°55'S	044°20'E	—	12:13:47.7	273	143	24	13:27:34.9			13:29:43.9							13:29:36.9	4	243	11	301	1.008	1.000	0.430	02m09s
Manja	21°26'S	044°20'E	—	12:13:47.7	273	144	22											13:27:58.5	184	64	12	301	0.998	0.999		
Marerano	21°23'S	044°52'E	—	12:14:40.4	272	143	25	13:28:08.2	104	344	13:28:42.8	265	145					13:28:25.3	184	65	11	301	1.000	1.000	0.024	00m34s
Midongy Sud	23°35'S	047°01'E	—	12:16:28.8	274	144	22	13:27:42.7	142	22	13:29:22.1	226	106					13:28:32.6	4	244	9	300	1.005	1.000	0.253	01m39s
Morombe	21°45'S	043°22'E	—	12:10:22.7	274	142	26	13:25:48.1	122	1	13:28:09.8	246	126					13:26:59.1	4	243	13	302	1.010	1.000	0.536	02m22s
Ranohira	22°29'S	045°24'E	—	12:14:48.9	275	143	24	13:26:52.6	107	347	13:29:23.0	261	141					13:28:08.0	4	244	10	301	1.015	1.000	0.780	02m30s
Ranomena	23°25'S	047°17'E	—	12:16:57.6	273	144	23	13:27:39.1	89	329	13:29:59.9	254	134					13:28:50.1	4	244	8	300	1.012	1.000	0.657	02m20s
Ranotsara Nord	22°48'S	046°36'E	—	12:16:23.5	273	143	24	13:27:31.8	89	329	13:30:02.1	279	160					13:28:47.2	184	64	9	301	1.017	1.000	0.907	02m30s
Sakaraha	22°55'S	044°32'E	—	12:13:13.5	275	143	25											13:27:10.9	4	243	11	301	0.993	0.996		
Tangainony	22°42'S	047°45'E	—	12:18:05.2	273	144	22	13:29:20.4	17	258	13:29:53.1	352	233					13:29:36.9	184	65	8	300	1.000	1.000	0.025	00m33s
Toamasina	18°10'S	049°23'E	—	12:23:44.9	265	143	22											13:33:07.7	185	71	8	298	0.855	0.825		
Toliara	23°21'S	043°40'E	—	12:11:37.1	276	143	25											13:28:28.2	4	242	12	301	0.972	0.972		
Tritriva	22°46'S	046°07'E	—	12:15:42.5	274	143	24	13:27:12.9	104	344	13:29:43.0	265	145					13:28:28.2	4	244	10	301	1.016	1.000	0.834	02m30s
Vangaindrano	23°21'S	047°36'E	—	12:17:26.6	274	144	22	13:27:51.0	97	337	13:30:19.0	271	152					13:29:05.2	4	244	8	300	1.018	1.000	0.954	02m28s
Vohipeno	22°22'S	047°51'E	—	12:18:27.2	272	144	22											13:29:53.6	184	66	8	300	0.990	0.993		
Vondrozo	22°49'S	047°20'E	—	12:17:25.6	273	144	22	13:28:13.2	62	302	13:30:18.7	307	188					13:29:16.2	184	65	8	300	1.009	1.000	0.461	02m06s
Zazafotsy	22°13'S	046°26'E	—	12:16:31.6	273	143	23	13:28:18.2	40	281	13:29:47.4	328	209					13:29:03.0	184	65	10	300	1.004	1.000	0.191	01m29s

69

TABLE 13
LOCAL CIRCUMSTANCES FOR MOZAMBIQUE
TOTAL SOLAR ECLIPSE OF 2001 JUNE 21

Location Name	Latitude	Longitude	Elev. m	First Contact U.T. h m s	P °	V °	Alt °	Second Contact U.T. h m s	P °	V °	Third Contact U.T. h m s	P °	V °	Fourth Contact U.T. h m s	P °	V °	Alt °	Maximum Eclipse U.T. h m s	P °	V °	Alt °	Azm °	Eclip. Mag.	Eclip. Obs.	Umbral Depth	Umbral Durat.
MOZAMBIQUE																										
Beira	19°49'S	034°52'E	9	11:56:34.0	275	135	36	—			—			14:29:35.2	91	335	8	13:18:58.7	3	239	22	307	0.964	0.964	0.229	
Campo	17°44'S	036°21'E	—	12:00:41.5	271	135	36	13:21:00.2	43	282	13:22:57.2	324	203	14:31:33.9	95	343	7	13:21:58.9	184	62	21	305	1.005	1.000		01m57s
Cataxa	15°58'S	033°12'E	—	11:54:10.6	270	131	40	—			—			14:30:31.1	96	343	11	13:18:35.3	183	61	25	306	0.993	0.996	0.851	
Changara	16°54'S	033°14'E	—	11:53:52.5	271	132	39	13:16:36.6	85	321	13:19:49.5	282	159	14:30:09.5	94	341	10	13:18:13.4	183	60	25	307	1.006	1.000		03m13s
Chemba	17°08'S	034°52'E	—	11:57:37.8	271	134	38	—			—			14:31:02.6	95	342	9	13:20:22.1	184	61	23	306	0.993	0.995	0.275	
Chicoa	15°37'S	032°24'E	—	11:52:20.6	270	130	41	13:16:16.7	47	285	13:21:27.1	320	198	14:30:07.6	96	343	12	13:17:33.9	183	61	26	307	1.014	1.000		02m10s
Chinde	18°37'S	036°24'E	—	12:00:24.1	273	136	35	13:20:09.8	114	351	13:23:01.4	253	131	14:31:09.9	94	340	7	13:21:35.9	4	241	21	305	1.010	1.000		02m52s
Chioco	16°25'S	032°50'E	—	11:53:05.5	271	131	40	13:16:29.4	60	296	13:19:13.5	307	184	14:30:06.3	95	342	11	13:17:51.7	183	60	26	307	1.007	1.000		02m44s
Conceiãao	18°45'S	036°10'E	—	11:59:50.3	273	136	35	13:20:04.3	134	11	13:22:23.9	233	111	14:30:58.3	94	340	9	13:21:14.4	4	241	21	305	0.999	1.000	0.350	02m20s
Doa	16°44'S	034°32'E	—	11:57:01.9	271	133	38	—			—			14:31:01.2	96	343	9	13:20:06.4	184	62	24	306	—	1.000	0.503	
Dona Ana	17°25'S	035°07'E	—	11:58:04.7	272	134	37	13:19:12.0	64	301	13:21:55.3	303	181	14:31:03.5	95	342	8	13:20:33.9	184	61	23	306	1.011	1.000		02m43s
Inhaminga	18°24'S	035°00'E	—	11:57:24.2	273	134	37	13:19:03.0	150	26	13:19:28.9	217	94	14:30:47.5	94	341	8	13:19:55.5	2	240	22	306	1.004	1.000	0.169	01m45s
Lacerdónia	18°01'S	035°30'E	—	11:58:41.1	272	134	36	13:19:12.9	97	335	13:22:18.9	270	147	14:30:59.3	94	341	8	13:20:46.2	4	241	22	306	1.020	1.000		03m06s
Luabo	18°30'S	036°10'E	—	11:59:56.8	273	136	35	13:19:55.0	113	350	13:22:48.8	255	132	14:31:06.2	94	340	7	13:21:22.2	4	241	21	305	1.014	1.000		02m54s
Macosa	17°52'S	033°56'E	—	11:55:30.5	273	133	38	—			—			14:30:08.4	93	339	9	13:18:44.9	3	240	24	307	1.006	0.996	0.286	02m14s
Macuze	17°42'S	037°11'E	—	12:02:30.6	271	136	35	13:17:37.5	139	15	13:19:15.9	228	104	14:31:58.4	96	344	6	13:23:00.2	184	63	20	304	0.994	0.996		
Mágoé	15°48'S	031°43'E	—	11:50:32.6	270	129	42	13:15:20.9	44	280	13:17:33.2	322	199	14:29:35.4	95	342	12	13:16:44.4	183	60	27	307	1.005	1.000	0.245	02m12s
Mandié	16°30'S	033°30'E	—	11:54:40.8	271	132	39	13:17:48.2	41	278	13:19:45.4	326	204	14:30:29.4	95	342	10	13:18:47.0	183	61	25	306	1.004	1.000	0.202	01m57s
Maputo (Lourenç...)	25°58'S	032°35'E	59	11:50:03.0	284	135	33	—			—			14:22:07.9	80	317	8	13:11:31.0	2	230	21	311	0.784	0.737	0.761	
Marromeu	18°20'S	035°56'E	—	11:59:30.5	273	135	36	13:19:39.8	107	345	13:22:39.2	260	138	14:31:03.9	94	340	7	13:21:09.8	4	241	21	305	1.016	1.000		02m59s
Massara	18°20'S	034°09'E	—	11:55:29.3	273	133	38	13:20:28.8	84	322	13:23:28.9	283	162	14:30:01.8	93	338	9	13:18:48.8	3	240	23	307	0.996	0.998	0.835	03m00s
Micaúne	18°18'S	036°35'E	—	12:00:56.3	272	136	35	—			—			14:31:25.0	95	341	7	13:21:59.2	184	62	21	305	1.017	1.000		03m04s
Moatize	16°08'S	033°45'E	—	11:53:45.8	270	132	39	13:12:44.0	66	301	13:15:48.0	300	176	14:28:37.7	95	341	10	13:14:16.3	183	58	25	309	0.991	0.994	0.540	03m05s
Molumbo	15°27'S	030°15'E	—	11:46:51.8	270	127	43	13:19:32.5	88	325	13:22:37.2	280	158	14:31:07.9	94	341	14	13:21:05.1	184	61	22	305	1.012	1.000		01m45s
Mopeia Velha	17°59'S	035°44'E	—	11:59:13.1	272	135	36	13:20:20.3	38	276	13:22:05.0	330	208	14:31:31.0	95	343	8	13:21:05.8	184	61	22	305	1.004	1.000	0.172	03m17s
Morrumbala	17°22'S	035°36'E	—	11:59:11.7	271	134	37	13:14:20.8	81	316	13:17:38.1	285	162	14:29:18.3	94	341	12	13:15:59.8	183	59	27	308	1.017	1.000	0.786	02m17s
Mucumbura	16°09'S	031°31'E	—	11:49:54.0	271	129	42	13:13:30.9	53	291	13:15:48.0	315	194	14:29:39.4	94	340	10	13:18:29.2	3	240	24	307	1.020	1.000	0.344	03m13s
Mucupia	18°01'S	036°48'E	—	12:03:07.8	271	136	35	13:16:52.3	98	335	13:20:05.5	269	145	14:30:12.2	94	340	8	13:18:47.0	4	241	21	305	0.864	0.838	0.916	
Mungári	17°12'S	033°31'E	—	11:54:26.2	272	132	39	—			—			14:33:53.4	104	356	4	13:28:04.4	184	69	18	301				
Nacala	14°34'S	040°41'E	—	12:11:36.5	265	137	33	—			—			14:33:53.4	104	356	4	13:28:04.4	184	69	18	301	0.864	0.838		
Namacurra	17°29'S	037°01'E	—	12:02:15.5	271	136	35	—			—			14:31:59.3	96	344	7	13:22:54.6	184	63	21	304	0.990	0.993	0.273	02m11s
Nampula	15°07'S	039°15'E	—	12:08:18.9	266	136	34	—			—			14:33:28.7	102	353	5	13:26:25.4	184	67	20	302	0.898	0.880	0.142	
Nhamacolomo	18°05'S	034°26'E	—	11:56:14.3	273	133	38	13:18:14.1	140	16	13:20:24.7	227	104	14:30:20.3	93	339	9	13:19:19.6	3	240	23	306	1.006	1.000	0.610	01m33s
Quelimane	17°53'S	036°51'E	—	12:01:42.4	271	136	35	13:21:44.1	35	273	13:23:17.5	333	212	14:31:44.4	95	343	7	13:22:37.5	184	62	21	305	1.003	1.000	0.119	02m54s
Sena	17°27'S	035°00'E	—	11:57:47.9	272	134	37	13:18:56.6	71	308	13:21:50.3	297	174	14:30:58.7	95	342	9	13:20:23.8	184	61	23	306	1.013	1.000		01m31s
Tambara	16°45'S	034°15'E	—	11:56:21.6	271	133	39	13:18:57.4	32	269	13:20:28.0	335	213	14:30:51.0	95	342	10	13:19:42.9	184	61	24	306	1.003	1.000		
Tete	16°13'S	033°35'E	—	11:55:00.4	270	132	40	—			—			14:30:39.7	96	343	10	13:19:01.5	183	61	25	306	0.996	0.998		
Vila Fontes	17°50'S	035°21'E	—	11:58:25.6	272	134	37	13:19:06.6	89	326	13:22:12.9	279	156	14:30:59.7	94	341	8	13:20:40.1	184	61	22	306	1.019	1.000		03m06s
Vila Gouveia	18°03'S	033°11'E	—	11:48:30.4	270	127	43	—			—			14:29:33.2	93	339	11	13:17:35.8	3	239	24	307	0.998	0.996	0.913	
Zâmbuè	15°10'S	030°50'E	—	11:47:15.0	271	127	43	13:12:52.1	72	307	13:16:05.1	294	169	14:29:10.9	95	342	13	13:15:19.8	183	59	28	308	0.998	0.999		03m13s
Zumbo	15°36'S	030°25'E	—	11:47:50.5	274	135	36	—			—			14:28:42.0	95	342	14	13:14:28.9	183	58	29	308	1.014	1.000	0.645	
Zune	18°59'S	035°18'E	—	11:57:50.5	274	135	36	—			13:16:05.1	294	169	14:30:21.7	93	338	8	13:20:00.7	3	240	22	306	0.991	0.994		

TABLE 14
LOCAL CIRCUMSTANCES FOR INDIAN OCEAN
TOTAL SOLAR ECLIPSE OF 2001 JUNE 21

Location Name	Latitude	Longitude	Elev. m	First Contact U.T. h m s	P °	V °	Alt °	Second Contact U.T. h m s	P °	V °	Third Contact U.T. h m s	P °	V °	Fourth Contact U.T. h m s	P °	V °	Alt °	Maximum Eclipse U.T. h m s	P °	V °	Alt °	Azm °	Eclip. Mag.	Eclip. Obs.
COMOROS																								
Moroni	11°41'S	043°16'E	—	12:18:55.4	259	137	32	—			—			14:33:54.9	110	6	3	13:31:14.7	185	74	17	299	0.749	0.692
MAURITIUS																								
Port Louis	20°10'S	057°30'E	55	12:31:44.9	263	144	13	—			—			—				13:35:16.9	185	73	1	295	0.808	0.765
REUNION																								
St.-Denis	20°52'S	055°28'E	936	12:29:00.5	265	144	15	—			—			—				13:34:21.6	185	71	2	296	0.854	0.823
SEYCHELLES																								
Victoria	04°38'S	055°27'E	5	12:48:01.2	236	132	19	—			—			—				13:38:36.9	185	87	8	294	0.378	0.265

TABLE 15
LOCAL CIRCUMSTANCES FOR ZAMBIA & ZIMBABWE
TOTAL SOLAR ECLIPSE OF 2001 JUNE 21

Location Name	Latitude	Longitude	Elev. (m)	First Contact U.T.	P°	V°	Alt°	Second Contact U.T.	P°	V°	Third Contact U.T.	P°	V°	Fourth Contact U.T.	P°	V°	Alt°	Maximum Eclipse U.T.	P°	V°	Alt°	Azm°	Eclip. Mag.	Eclip. Obs.		
ZAMBIA																										
Balovale	13°33'S	023°06'E	—	11:26:49.9	269	111	50	12:59:28.1	109	338	13:03:13.7	253	123	14:22:15.0	93	336	22	13:01:21.2	1	230	38	315	1.016	1.000	0.691	03m46s
Chavuma	13°05'S	022°40'E	—	11:25:36.8	268	109	51	12:58:34.2	86	315	13:02:32.9	276	145	14:21:53.4	93	336	23	13:00:33.4	181	50	38	315	1.021	1.000	0.914	03m59s
Chibwe	14°12'S	028°31'E	—	11:42:36.0	269	123	46							14:27:38.3	95	342	16	13:11:48.5	183	57	31	309	0.996	0.998		
Chilanga	15°34'S	028°17'E	—	11:41:30.0	271	123	45	13:09:25.9	130	3	13:12:16.1	235	108	14:26:54.4	96	338	16	13:10:51.3	2	236	31	310	1.009	1.000	0.391	02m50s
Chingola	12°32'S	027°52'E	—	11:41:25.0	267	121	48							14:27:29.6	98	345	18	13:11:12.6	183	58	33	309	0.957	0.955		
Chinyama Litapi	13°31'S	022°21'E	—	11:24:30.7	269	109	51	12:58:00.6	124	351	13:01:23.0	238	107	14:27:19.9	92	335	23	12:59:42.2	1	229	38	315	1.011	1.000	0.461	03m32s
Chisamba	14°58'S	028°23'E	—	11:41:57.8	270	123	45	13:09:30.5	83	316	13:13:02.1	283	157	14:27:14.9	94	339	16	13:11:16.7	183	57	31	310	1.018	1.000	0.827	03m32s
Chitokoloki	13°50'S	023°13'E	—	11:27:07.2	269	111	50	12:59:55.6	128	357	13:03:04.3	234	103	14:22:16.7	92	335	22	13:01:30.3	1	230	37	315	1.009	1.000	0.397	03m09s
Feira	15°37'S	030°25'E	—	11:47:14.7	271	127	43	13:12:50.9	73	309	13:16:05.4	293	168	14:28:41.6	92	340	14	13:14:28.5	183	58	29	308	1.015	1.000	0.666	03m15s
Ibwe Munyama	16°09'S	028°34'E	—	11:42:06.6	272	124	44							14:26:53.4	92	337	15	13:11:06.4	3	236	30	310	0.996	0.998		
Kabwe (Broken H...	14°27'S	028°27'E	—	11:42:19.4	269	123	46	13:10:36.9	36	270	13:12:34.2	330	204	14:27:29.9	95	341	16	13:11:35.8	183	57	31	310	1.004	1.000	0.162	01m57s
Kafue	15°47'S	028°11'E	—	11:41:09.6	271	123	45	13:09:52.4	159	32	13:11:17.2	206	79	14:26:42.9	93	337	16	13:10:35.0	3	235	31	311	1.002	1.000	0.081	01m25s
Kambanga	13°23'S	023°03'E	—	11:26:43.3	268	111	51	12:59:20.1	99	327	13:03:15.7	264	133	14:22:15.1	93	336	22	13:01:18.3	1	230	38	315	1.020	1.000	0.870	03m56s
Kangombe	14°03'S	023°40'E	—	11:28:26.7	269	113	50	13:00:59.0	135	4	13:03:47.9	227	97	14:22:43.7	92	335	21	13:02:23.7	1	231	37	314	1.007	1.000	0.307	02m49s
Kasempa	13°27'S	025°50'E	—	11:35:08.3	268	117	49	13:06:12.8	29	261	13:07:55.0	335	208	14:25:20.5	95	340	19	13:07:04.1	182	55	35	312	1.002	1.000	0.108	01m42s
Kitwe	12°49'S	028°13'E	—	11:42:17.8	267	122	47							14:27:44.5	97	345	17	13:11:45.0	183	59	32	309	0.961	0.961		
Lakulu	14°22'S	023°17'E	—	11:27:11.9	270	112	50							14:22:07.8	94	334	22	13:01:26.7	1	230	37	315	0.995	0.998		
Luanshya	13°08'S	028°24'E	—	11:42:41.1	267	122	47							14:27:49.8	97	344	17	13:11:58.4	183	58	32	309	0.968	0.969		
Lukulu	14°59'S	023°12'E	—	11:26:55.9	270	112	50							14:22:00.3	93	334	22	13:01:14.7	1	230	37	315	0.993	0.996		
Lusaka	15°25'S	028°17'E	1277	11:41:33.5	271	123	45	13:09:19.3	118	351	13:12:32.8	247	121	14:26:59.8	93	338	16	13:10:56.4	2	236	31	310	1.013	1.000	0.567	03m14s
Lusongwa	12°58'S	024°16'E	—	11:30:35.0	268	113	50	13:02:58.2	35	266	13:05:07.9	328	199	14:23:48.7	94	339	21	13:04:03.3	182	53	37	313	1.004	1.000	0.169	02m10s
Magoye	16°00'S	027°37'E	—	11:39:31.5	272	123	45							14:26:04.5	92	336	16	13:09:27.7	2	234	31	311	0.990	0.993		
Mankoya	14°47'S	024°48'E	—	11:31:41.2	270	116	48							14:23:43.0	92	335	20	13:09:29.4	2	232	35	313	0.997	0.999		
Mazabuka	15°51'S	027°46'E	—	11:39:59.0	271	123	45							14:26:17.5	92	336	16	13:09:48.1	2	235	31	311	0.996	0.998		
Mufulira	12°33'S	028°14'E	—	11:42:27.5	267	121	47							14:27:48.3	98	345	17	13:11:51.7	183	59	33	309	0.953	0.951		
Mukinge Hill	13°29'S	025°52'E	—	11:35:13.6	268	117	49	13:06:10.6	32	265	13:08:03.7	332	205	14:25:45.3	95	340	19	13:07:07.4	182	55	35	311	1.003	1.000	0.134	01m53s
Mulungushi	14°40'S	028°50'E	—	11:43:18.7	270	124	45	13:11:00.5	44	279	13:13:21.4	321	196	14:27:45.3	94	341	16	13:12:11.2	183	57	31	309	1.006	1.000	0.249	02m21s
Mumbwa	14°59'S	027°04'E	—	11:38:15.1	270	121	46	13:07:12.7	118	350	13:10:30.6	246	119	14:24:00.6	93	338	18	13:08:51.9	183	57	32	311	1.013	1.000	0.564	03m18s
Mushima	14°13'S	025°05'E	—	11:32:40.6	269	116	49	13:03:29.6	112	342	13:07:04.4	252	123	14:24:16.5	93	337	20	13:05:17.4	2	233	32	313	1.015	1.000	0.659	03m35s
Ndola	12°58'S	028°38'E	—	11:43:24.7	267	122	47							14:28:03.7	97	345	17	13:12:26.3	183	59	32	309	0.961	0.960		
Ngwerere	15°18'S	028°20'E	—	11:41:43.1	271	123	45	13:09:19.3	108	341	13:12:46.6	257	131	14:27:04.1	93	339	16	13:11:03.3	183	57	31	310	1.016	1.000	0.741	03m27s
Nyakulenga	13°03'S	023°29'E	—	11:28:09.3	268	111	51	13:00:34.7	66	296	13:04:08.2	297	167	14:22:53.2	92	338	17	13:02:53.2	181	51	37	314	1.013	1.000	0.572	03m34s
Old Mkushi	14°22'S	029°22'E	—	11:44:53.1	269	125	45							14:28:18.1	96	342	15	13:13:13.2	183	58	30	309	0.992	0.995		
Rufunsa	15°05'S	029°40'E	—	11:45:26.2	270	125	44	13:12:05.5	54	289	13:14:48.0	312	188	14:28:18.0	95	341	15	13:13:27.1	183	58	30	309	1.008	1.000	0.368	02m42s
Sikelenge	14°50'S	024°14'E	—	11:29:58.7	270	115	49							14:23:02.7	91	334	20	13:03:17.5	1	231	36	314	0.991	0.994		
ZIMBABWE																										
Bindura	17°19'S	031°20'E	—	11:49:02.0	273	129	41							14:28:36.7	92	337	12	13:15:10.7	3	238	27	308	0.993	0.995		
Bradley Institu...	17°02'S	031°27'E	1343	11:49:25.2	272	127	41	13:14:51.3	161	36	13:16:07.6	205	80	14:28:50.8	93	338	12	13:15:29.7	3	238	27	308	1.002	1.000	0.075	01m16s
Bulawayo	20°09'S	028°36'E	—	11:41:23.3	277	127	41							14:24:17.3	86	327	14	13:09:07.5	2	231	28	312	0.890	0.871		
Chipuriro	16°39'S	030°42'E	—	11:47:37.8	272	128	42	13:13:29.4	147	22	13:15:28.2	218	93	14:28:27.4	93	338	13	13:14:29.0	3	238	28	309	1.004	1.000	0.186	01m59s
Chirundu	15°59'S	028°54'E	—	11:43:04.0	273	125	44	13:10:48.0	150	23	13:12:43.3	216	89	14:27:16.2	92	336	12	13:11:45.7	3	236	30	310	1.004	1.000	0.161	01m55s
Chitungwiza	17°45'S	031°16'E	—	11:48:43.7	273	130	41							14:28:19.4	92	336	12	13:14:30.2	3	237	27	309	0.980	0.983		
Harare (Salisbu...	17°50'S	031°03'E	1472	11:48:10.6	274	129	41							14:28:07.8	91	336	12	13:14:30.2	3	239	27	309	0.976	0.978		
Makaha	17°17'S	032°37'E	—	11:52:13.1	272	131	41	13:15:54.1	134	10	13:18:22.9	232	108	14:29:33.9	93	339	11	13:17:16.2	3	238	26	307	1.007	1.000	0.345	02m29s
Miami	16°40'S	030°46'E	—	11:45:11.1	272	131	43							14:27:41.1	92	337	14	13:12:56.5	3	236	29	310	0.994	0.996		
Mtoko	17°24'S	032°13'E	—	11:51:12.5	273	131	40							14:29:13.4	93	338	11	13:16:29.6	3	237	26	308	1.000	1.000		
Rusambo	16°35'S	032°12'E	—	11:51:27.7	271	130	41	13:15:11.1	92	328	13:18:30.1	274	150	14:29:36.2	94	340	12	13:16:50.9	183	59	26	307	1.021	1.000	0.985	03m19s
Shamva	17°18'S	031°34'E	—	11:49:37.5	273	130	41							14:28:47.9	93	338	12	13:15:33.0	3	238	26	308	0.996	0.998		

TABLE 16
SOLAR ECLIPSES OF SAROS SERIES 127

First Eclipse: 0991 Oct 10   Duration of Series: 1460.44 yrs.
Last Eclipse: 2452 Mar 21   Number of Eclipses: 82

Saros Summary: Partial: 40   Annular: 0   Total: 42   Hybrid: 0

Date	Eclipse Type	Gamma	Mag./ Width	Center Durat.		Date	Eclipse Type	Gamma	Mag./ Width	Center Durat.
0991 Oct 10	Pb	1.536	0.034			1712 Dec 28	T	0.034	155	04m15s
1009 Oct 20	P	1.511	0.077			1731 Jan 08	Tm	0.031	155	04m10s
1027 Nov 01	P	1.492	0.110			1749 Jan 18	T	0.026	155	04m07s
1045 Nov 11	P	1.478	0.134			1767 Jan 30	T	0.019	157	04m06s
1063 Nov 22	P	1.467	0.153			1785 Feb 09	T	0.008	159	04m07s
1081 Dec 03	P	1.460	0.166			1803 Feb 21	T	-0.008	163	04m09s
1099 Dec 14	P	1.453	0.177			1821 Mar 04	T	-0.028	168	04m14s
1117 Dec 25	P	1.446	0.188			1839 Mar 15	T	-0.056	173	04m20s
1136 Jan 05	P	1.437	0.202			1857 Mar 25	T	-0.089	177	04m28s
1154 Jan 15	P	1.427	0.220			1875 Apr 06	T	-0.129	182	04m37s
1172 Jan 27	P	1.411	0.246			1893 Apr 16	T	-0.176	186	04m47s
1190 Feb 06	P	1.391	0.281			1911 Apr 28	T	-0.230	190	04m57s
1208 Feb 17	P	1.365	0.326			1929 May 09	T	-0.289	193	05m07s
1226 Feb 28	P	1.334	0.382			1947 May 20	T	-0.353	196	05m13s
1244 Mar 10	P	1.296	0.453			1965 May 30	T	-0.423	198	05m15s
1262 Mar 21	P	1.251	0.534			1983 Jun 11	T	-0.495	199	05m11s
1280 Apr 01	P	1.200	0.630			2001 Jun 21	T	-0.570	200	04m57s
1298 Apr 12	P	1.144	0.736			2019 Jul 02	T	-0.646	201	04m33s
1316 Apr 22	P	1.080	0.857			2037 Jul 13	T	-0.724	201	03m58s
1334 May 04	P	1.014	0.984			2055 Jul 24	T	-0.801	202	03m17s
1352 May 14	T	0.943	438	02m18s		2073 Aug 03	T	-0.876	206	02m29s
1370 May 25	T	0.870	338	02m51s		2091 Aug 15	T	-0.949	236	01m38s
1388 Jun 04	T	0.794	302	03m20s		2109 Aug 26	P	-1.018	0.967	
1406 Jun 16	T	0.718	283	03m48s		2127 Sep 06	P	-1.082	0.846	
1424 Jun 26	T	0.642	270	04m14s		2145 Sep 16	P	-1.140	0.737	
1442 Jul 07	T	0.567	260	04m39s		2163 Sep 28	P	-1.194	0.638	
1460 Jul 18	T	0.495	252	05m00s		2181 Oct 08	P	-1.240	0.553	
1478 Jul 29	T	0.426	244	05m18s		2199 Oct 19	P	-1.281	0.479	
1496 Aug 08	T	0.362	236	05m30s		2217 Oct 31	P	-1.315	0.419	
1514 Aug 20	T	0.303	228	05m38s		2235 Nov 11	P	-1.344	0.369	
1532 Aug 30	T	0.250	220	05m40s		2253 Nov 21	P	-1.366	0.330	
1550 Sep 10	T	0.203	212	05m37s		2271 Dec 03	P	-1.384	0.300	
1568 Sep 21	T	0.162	204	05m31s		2289 Dec 13	P	-1.397	0.277	
1586 Oct 12	T	0.127	196	05m23s		2307 Dec 25	P	-1.408	0.259	
1604 Oct 22	T	0.100	188	05m12s		2326 Jan 05	P	-1.417	0.245	
1622 Nov 03	T	0.079	180	05m01s		2344 Jan 16	P	-1.427	0.229	
1640 Nov 13	T	0.062	173	04m50s		2362 Jan 27	P	-1.436	0.213	
1658 Nov 24	T	0.051	167	04m40s		2380 Feb 07	P	-1.449	0.191	
1676 Dec 05	T	0.043	162	04m30s		2398 Feb 17	P	-1.464	0.165	
1694 Dec 16	T	0.039	158	04m22s		2416 Feb 29	P	-1.486	0.127	
						2434 Mar 11	P	-1.512	0.083	
						2452 Mar 21	Pe	-1.545	0.026	

Eclipse Type: P - Partial   Pb - Partial Eclipse (Saros Series Begins)
              T - Total   Pe - Partial Eclipse (Saros Series Ends)
                                 Tm - Middle eclipse of Saros series.

Note: Mag./Width column gives either the eclipse magnitude (for partial eclipses)
       or the umbral path width in kilometers (for total eclipses).

Table 17a
Meteorological Data for Southern Africa in June
Total Solar Eclipse of 2001 June 21

Station	Latitude (S)	Longitude (E)	mean daily sunshine hours	Percent of possible sunshine	clear sky	scattered cloud	broken cloud	overcast cloud	obscured sky	Monthly Pcpn (mm)	Mean Tmax (°C)	Mean Tmin (°C)	Prevailing Wind (knots)	Percent hours in June with less than 3/10th cloud and visibility > 4 km	Mean Monthly cloud amount	Fog	TRW	Rain	Smoke or Haze	Blowing Sand or Dust
St Helena																				
St Helena	-15.93	-5.67			0.2	19.6	74.4	4.2	1.6	0.7	18.9	17.2	ESE14	4		13.5	0	54.9	2.2	0.2
Namibia																				
Grootfontein	-19.60	18.12			67.2	25.4	7.0	0.4	0.0	0.0	22	7	E 8	84		0	0	0	7.3	1.2
Rundu	-17.92	19.77			73.0	19.8	6.9	0.0	0.2	0.0	24	8	E 9	81		0	0	0	1.0	1.0
Congo																				
Pointe-Noire	-4.82	11.90			0.3	36.8	38.6	24.3	0.0	0.0	26	21	SW 7	6		0	0	0.3	0	0
Brazzaville	-4.25	15.25			6.5	23.1	41.1	29.3	0.0	5.1	27	21	W 4	13		3.1	0	0.5	5.7	0.3
Angola																				
Nova Lisboa (Huambo)	-12.80	15.75	9.0	0.79						0.0	24	8	E 6	92	0.13	0				
Silva Porto (Bie)	-12.38	16.95								0.0	23	7		67						
Luanda New	-8.85	13.23								1.3	25	20		63						
Lubango	-14.93	13.57	9.5	0.84						1	24	8	E 6		0.13					
Zambia																				
Lusaka	-15.32	28.45	8.9	0.79	18.8	43.1	36.7	1.4	0.0	0	23	10	ESE 9	37	0.25	0	0	0.2	29.1	0
Kasama	-10.22	31.13	9.9	0.86						0.4	25	10	ESE 12		0.30					
Mongu	-15.25	23.17	9.9	0.88						0.7	27	10	E 12		0.14					
Kabwe	-14.45	28.45	9.3	0.82						0.1	24	9			0.28					
Kasempa	-13.45	25.83	9.3	0.82						0	25	6	E 5		0.25					
Livingstone	-17.82	29.82	9.7	0.88						0.6	26	7	E 6		0.19		0			
Chipata	-13.57	32.58	8.6	0.76						1.1	25	12	SE 9		0.33					
Ndola	-13.00	28.65	9.2	0.81						0.7	25	8	ESE 7		0.26			0		
Zimbabwe																				
Beitbridge	-22.22	-30.00	8.1	0.75						3.1	25.5	8.3			0.25					
Bualwayo	-20.15	-28.62	8.6	0.83	29.4	49.2	21.4	0.0	0.0	1.9	21.4	7.3	E 9	55	0.24	0	0.4	1.2	1.6	0
Chipinge	-20.20	-32.62	7.6	0.70						16.7	20.2	10.3			0.35					
Mount Darwin	-16.78	-31.58	8.2	0.74						0.9	24.2	6.8			0.31					
Gokwe	-18.22	-28.93	9.2	0.83						1.2	22.6	8.9			0.21					
Gweru Thornhill	-19.45	-29.85	8.7	0.80	32.3	46.5	19.7	1.6	0.0	1.9	20.4	4.8	ESE 9	59	0.28	0	0	0.4	1.2	0
Harare Kutsaga	-17.92	-31.13	8.9	0.80	23.0	46.0	30.6	0.4	0.0	2.2	21.4	5.4	E 6	49	0.25	0	0.4	2.0	7.5	0.4
Kadoma	-18.32	-29.88	9.1	0.82						1.6	24.0	8.8			0.23					
Kariba	-16.52	-28.88	9.1	0.82	17.0	59.4	23.6	0.0	0.0	1.0	26.5	11.5	W 6	48	0.24	0	0	0.9	4.6	0
Karoi	-16.83	-29.62	8.5	0.76	15.5	50.1	33.5	0.8	0.0	1.7	22.2	8.7	E 5	46	0.28	0	0	1.1	6.2	0
Marondera	-18.18	-31.47	8.4	0.76						5.4	19.5	6.1			0.30					
Masvingo	-20.07	-30.87	8.0	0.73						5.7	21.7	6.0			0.30					
Nyanga	-18.28	-32.75	8.0	0.72						17.4	16.4	5.4			0.33					
Victoria Falls	-18.10	-25.85	9.7	0.88						0.6	24.7	6.0			0.15					
Mutare (Umtali)	-18.97	32.67	9.1	0.83						9.0	21	9	ESE 5		0.25					
Malawi																				
Blantyre	-15.68	34.97	7.1	0.63	5.3	30.7	63.6	0.4	0.0	4.0	24	13	SSE3	18	0.38	0	0	0.4	4.2	0
Lilongwe	-13.78	33.77	7.7	0.68						1.0	22	10	E 8		0.38					
Zomba	-15.38	35.32	5.3	0.47						13.0	22	13	E4		0.50					
Mlanje	-16.08	35.63	5.9	0.53						64.0	24	12	SE4		0.50					

73

Table 17b
Meteorological Data for Southern Africa in June
Total Solar Eclipse of 2001 June 21

Station	Latitude (S)	Longitude (E)	mean daily sunshine hours	Percent of possible sunshine	clear sky	scattered cloud	broken cloud	overcast cloud	obscured sky	Monthly Pcpn (mm)	Mean Tmax (°C)	Mean Tmin (°C)	Prevailing Wind (knots)	Percent hours in June with less than 3/10th cloud and visibility > 4 km	Mean Monthly cloud amount	Fog	TRW	Rain	Smoke or Haze	Blowing Sand or Dust
						Percent Frequency of (at nearest hour to eclipse)											Percent of Observations at Eclipse Time in June with			
Mozambique																				
Quelimane	-17.88	36.88			4.3	45.7	46.2	3.8	0.0	64.5	26	18	S 7	15		0	0	8.9	4.1	0.5
Cuamba (Nova Freixo)	-14.80	36.87								2.0	28	12			0.38					
Beira	-19.80	34.90	7.4	0.68	10.1	65.2	20.5	4.1	0.0	40.6	24	19	S 9	33	0.13	0	0	4.4	14.3	0.8
Maputo	-25.92	32.57	8.2	0.78	31.0	42.8	24.0	2.2	0.0	17.8	24	15	N 7	46	0.25	0	0	3.1	20.9	1.4
Nampula	-15.10	39.28			1.7	27.4	63.8	7.1	0.0	20.3	24	18	S 7	5		0	0	4.0	9.0	0.2
Tete	-16.18	38.58	6.5	0.58						3.0	30	15	SE5		0.38					
Fingoe	-15.17	31.88								8.0	26	12	SE		0.38					
Madagascar																				
Antalaha	-15.00	50.33	8.1	0.72	0.8	49.1	50.1	0.0	0.0	162	26	19	SSE 12	14	0.63	0	0	3.6	0	0
Antananarivo	-18.80	47.48	7.1	0.64	0.7	52.8	44.1	2.4	0.0	7.6	20	12	E 7	22	0.50	0.2	0.7	9.4	0	0.2
Taolagnaro (Fort Dauphin)	-25.03	46.95	7.0	0.66	0.0	59.8	38.6	1.6	0.0	122	24	18	E 11	16	0.50	0	0	3.4	0.2	0
Antsir	-12.35	49.30			0.5	42.9	50.4	6.1	0.0	-	28	22	SE 14	14		0.3	0	12.8	0.5	0
Mahanoro	-19.83	48.80								-	24	19	SW 6							
Mahajanga	-15.67	46.33	9.9	0.88						3.0	31	19	ESE 12		0.25					
Maintirano	-18.05	44.03	9.8	0.89						4.0	27	18	SW 13		0.25					
Nossi Be	-13.33	48.28	9.4	0.83						49.0	29	19	NE 5		0.38					
Sambava Sud	-14.28	50.17			0.0	42.1	51.7	6.3	0.0	-	26	21	S 11	14		0.6	0.2	10.9	0	0.2
Toliara (Tulear)	-23.38	43.73	9.2	0.86						11.0	27	15	SSW 13		0.25					
Reunion																				
Europa Island	-22.32	40.33			7.6	74.4	17.4	0.6	0.0		25	21	SE 10	34		0.2	0	1.1	0	0.4
Saint Denis, Reunion Is.	-20.88	55.52			0.0	53.6	41.6	1.9	0.2		26	19	SE 10	13		0.0	0.3	8.7	0.2	0
Mayotte																				
Dzaoudzi	-12.80	45.28			0.2	78.0	21.0	0.9	0.0	8.9	28	24	SSE 10	10		0.2	0	2.4	0	0

Explanations
Percent of possible sunshine: mean hours of sunshine / mean hours of daylight in June
Scattered cloud: four tenths or less
Broken cloud: five to nine tenths
Tmax: Maximum temperature
Tmin: Minimum temperature
TRW: thundershowers

TABLE 18

35 MM FIELD OF VIEW AND SIZE OF SUN'S IMAGE FOR VARIOUS PHOTOGRAPHIC FOCAL LENGTHS

Focal Length	Field of View	Size of Sun
28 mm	49° x 74°	0.2 mm
35 mm	39° x 59°	0.3 mm
50 mm	27° x 40°	0.5 mm
105 mm	13° x 19°	1.0 mm
200 mm	7° x 10°	1.8 mm
400 mm	3.4° x 5.1°	3.7 mm
500 mm	2.7° x 4.1°	4.6 mm
1000 mm	1.4° x 2.1°	9.2 mm
1500 mm	0.9° x 1.4°	13.8 mm
2000 mm	0.7° x 1.0°	18.4 mm
2500 mm	0.6° x 0.8°	22.9 mm

Image Size of Sun (mm) = Focal Length (mm) / 109

TABLE 19

SOLAR ECLIPSE EXPOSURE GUIDE

ISO				f/Number					
25	1.4	2	2.8	4	5.6	8	11	16	22
50	2	2.8	4	5.6	8	11	16	22	32
100	2.8	4	5.6	8	11	16	22	32	44
200	4	5.6	8	11	16	22	32	44	64
400	5.6	8	11	16	22	32	44	64	88
800	8	11	16	22	32	44	64	88	128
1600	11	16	22	32	44	64	88	128	176

Subject	Q				Shutter Speed				
Solar Eclipse									
Partial[1] - 4.0 ND	11	—	—	—	1/4000	1/2000	1/1000	1/500	1/250	1/125
Partial[1] - 5.0 ND	8	1/4000	1/2000	1/1000	1/500	1/250	1/125	1/60	1/30	1/15
Baily's Beads[2]	11	—	—	—	1/4000	1/2000	1/1000	1/500	1/250	1/125
Chromosphere	10	—	—	1/4000	1/2000	1/1000	1/500	1/250	1/125	1/60
Prominences	9	—	1/4000	1/2000	1/1000	1/500	1/250	1/125	1/60	1/30
Corona - 0.1 Rs	7	1/2000	1/1000	1/500	1/250	1/125	1/60	1/30	1/15	1/8
Corona - 0.2 Rs[3]	5	1/500	1/250	1/125	1/60	1/30	1/15	1/8	1/4	1/2
Corona - 0.5 Rs	3	1/125	1/60	1/30	1/15	1/8	1/4	1/2	1 sec	2 sec
Corona - 1.0 Rs	1	1/30	1/15	1/8	1/4	1/2	1 sec	2 sec	4 sec	8 sec
Corona - 2.0 Rs	0	1/15	1/8	1/4	1/2	1 sec	2 sec	4 sec	8 sec	15 sec
Corona - 4.0 Rs	-1	1/8	1/4	1/2	1 sec	2 sec	4 sec	8 sec	15 sec	30 sec
Corona - 8.0 Rs	-3	1/2	1 sec	2 sec	4 sec	8 sec	15 sec	30 sec	1 min	2 min

Exposure Formula: $t = f^2 / (I \times 2^Q)$ where: t = exposure time (sec)
f = f/number or focal ratio
I = ISO film speed
Q = brightness exponent

Abbreviations: ND = Neutral Density Filter.
Rs = Solar Radii.

Notes: [1] Exposures for partial phases are also good for annular eclipses.
[2] Baily's Beads are extremely bright and change rapidly.
[3] This exposure also recommended for the 'Diamond Ring' effect.

F. Espenak – 1997 Feb

TOTAL SOLAR ECLIPSE OF 2001 JUNE 21

MAPS

Total Solar Eclipse of 2001 June 21

Map 1 - Coastal Angola

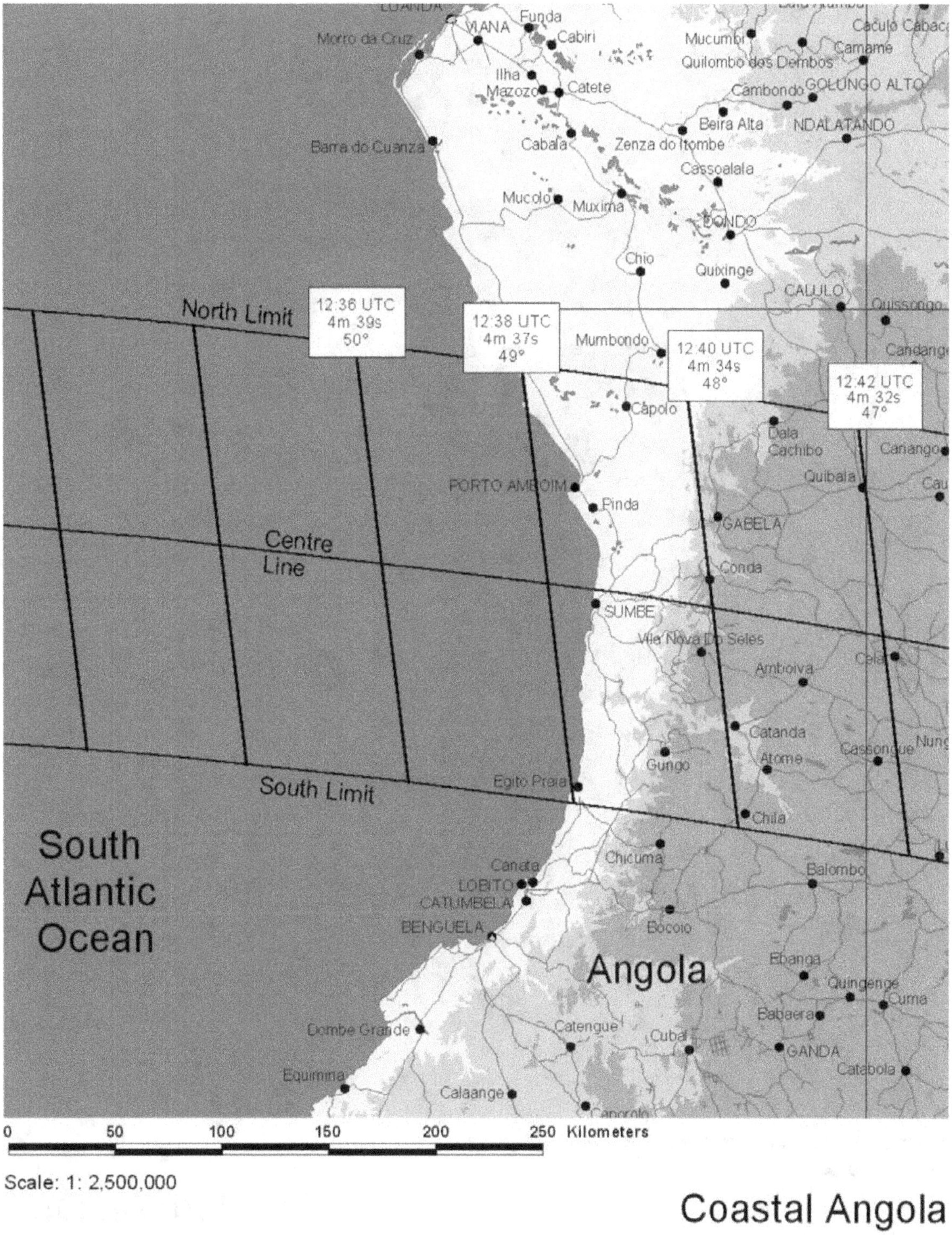

Coastal Angola

Scale: 1: 2,500,000

Total Solar Eclipse of 2001 June 21

Map 2 - Western Angola

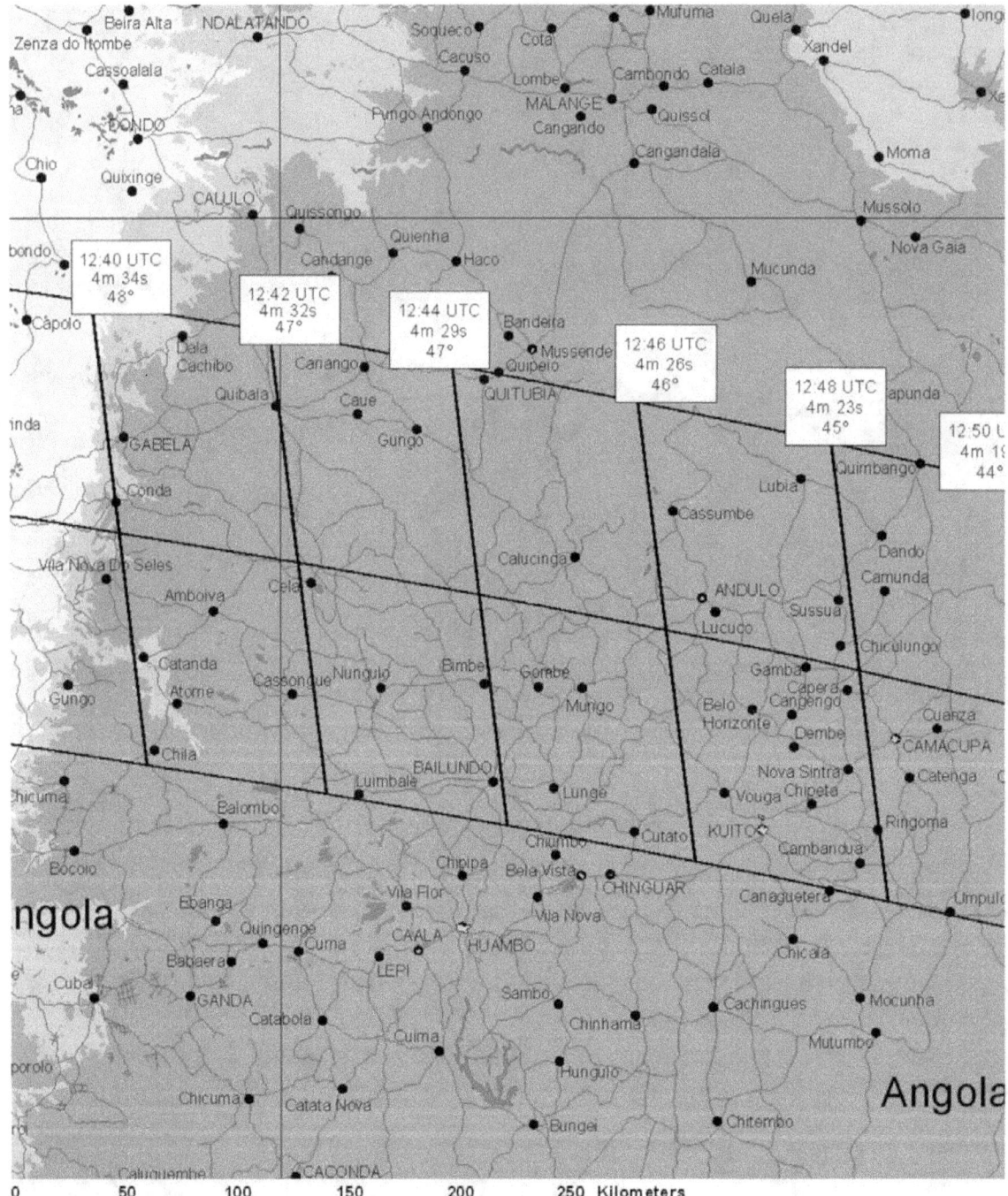

Scale: 1:2,500,000

Western Angola

Total Solar Eclipse of 2001 June 21

Map 3 - Angola

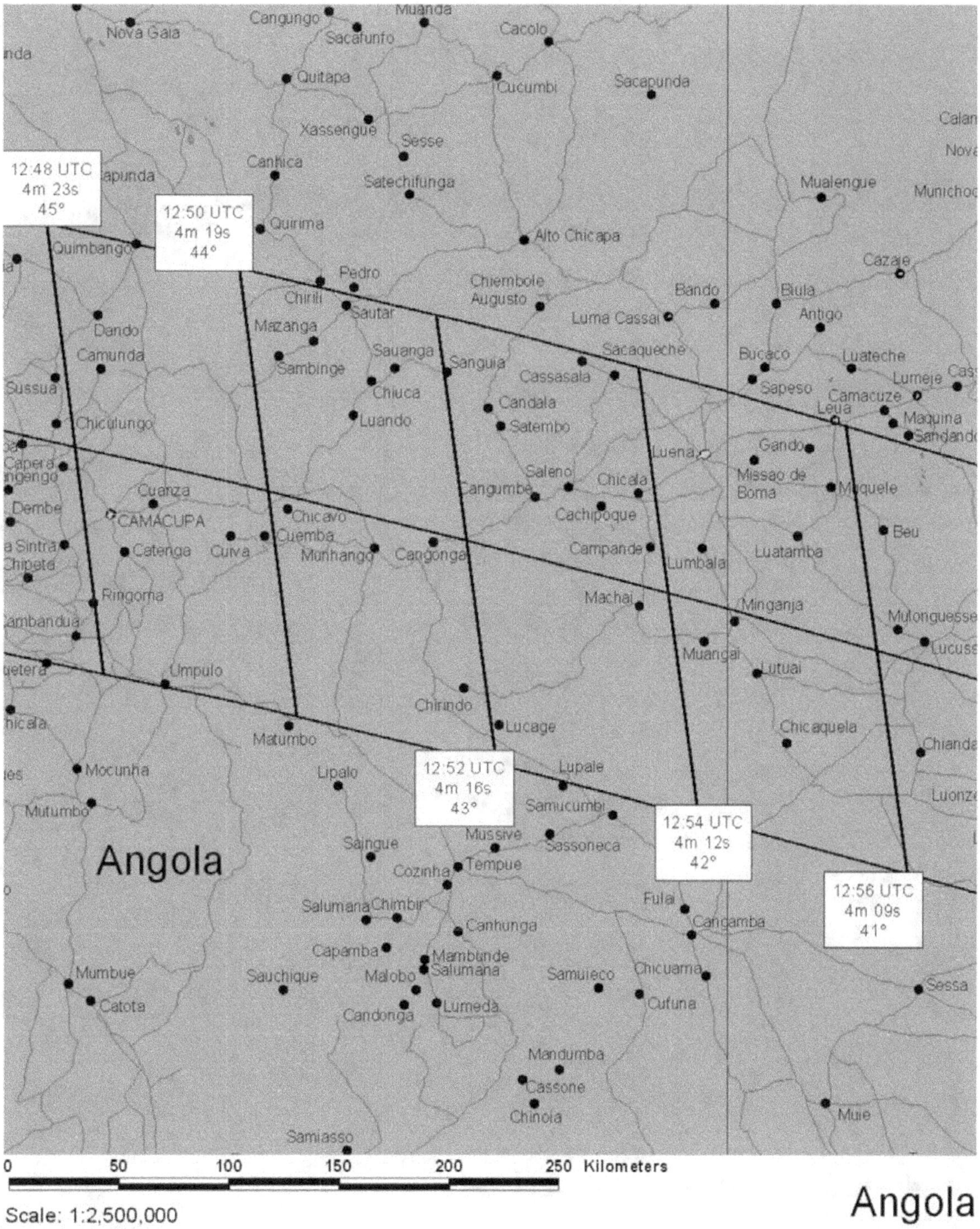

Total Solar Eclipse of 2001 June 21

Map 4 - Angola - Zambia

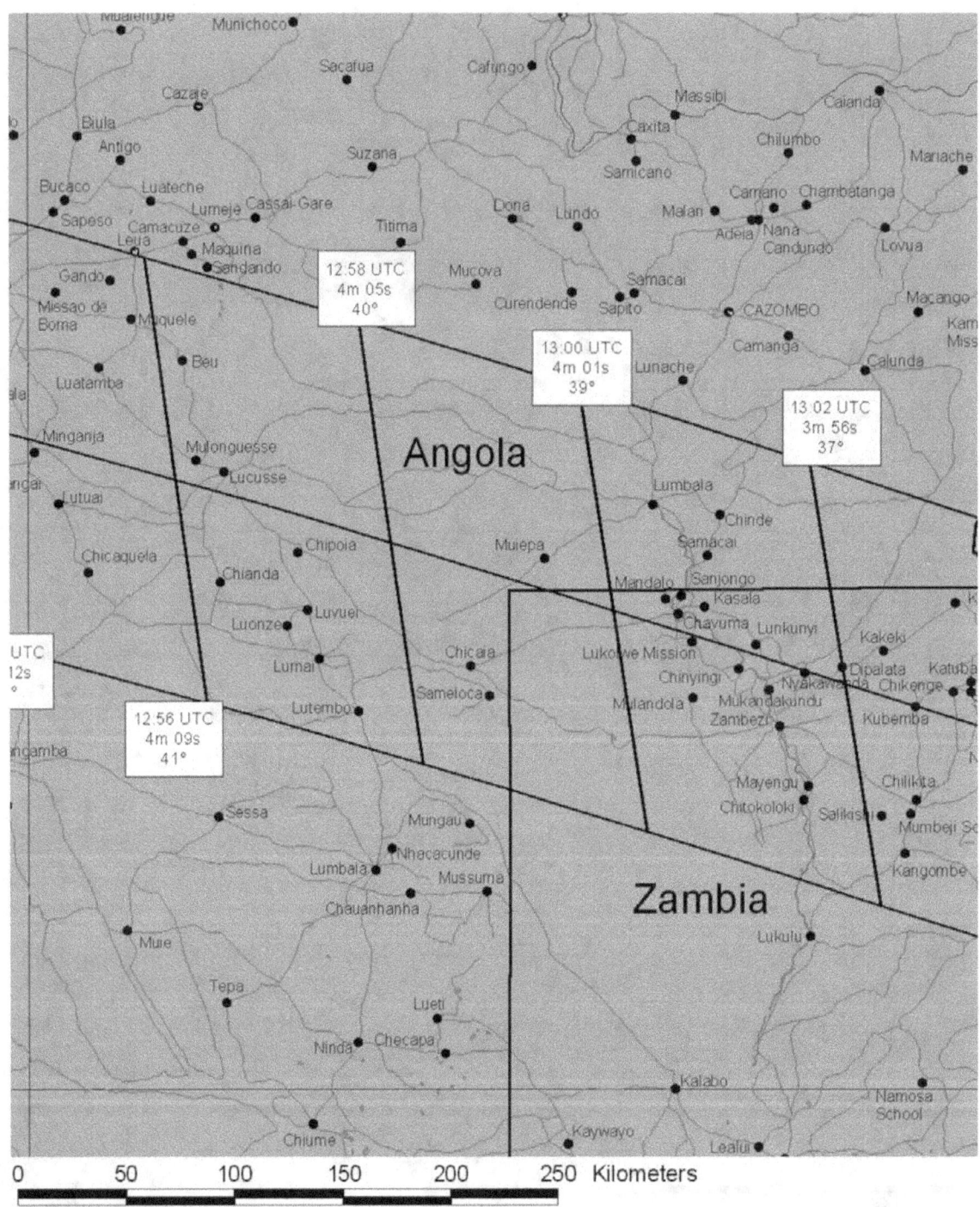

Scale: 1:2,500,000

Angola - Zambia

Total Solar Eclipse of 2001 June 21

Map 5 - Zambia

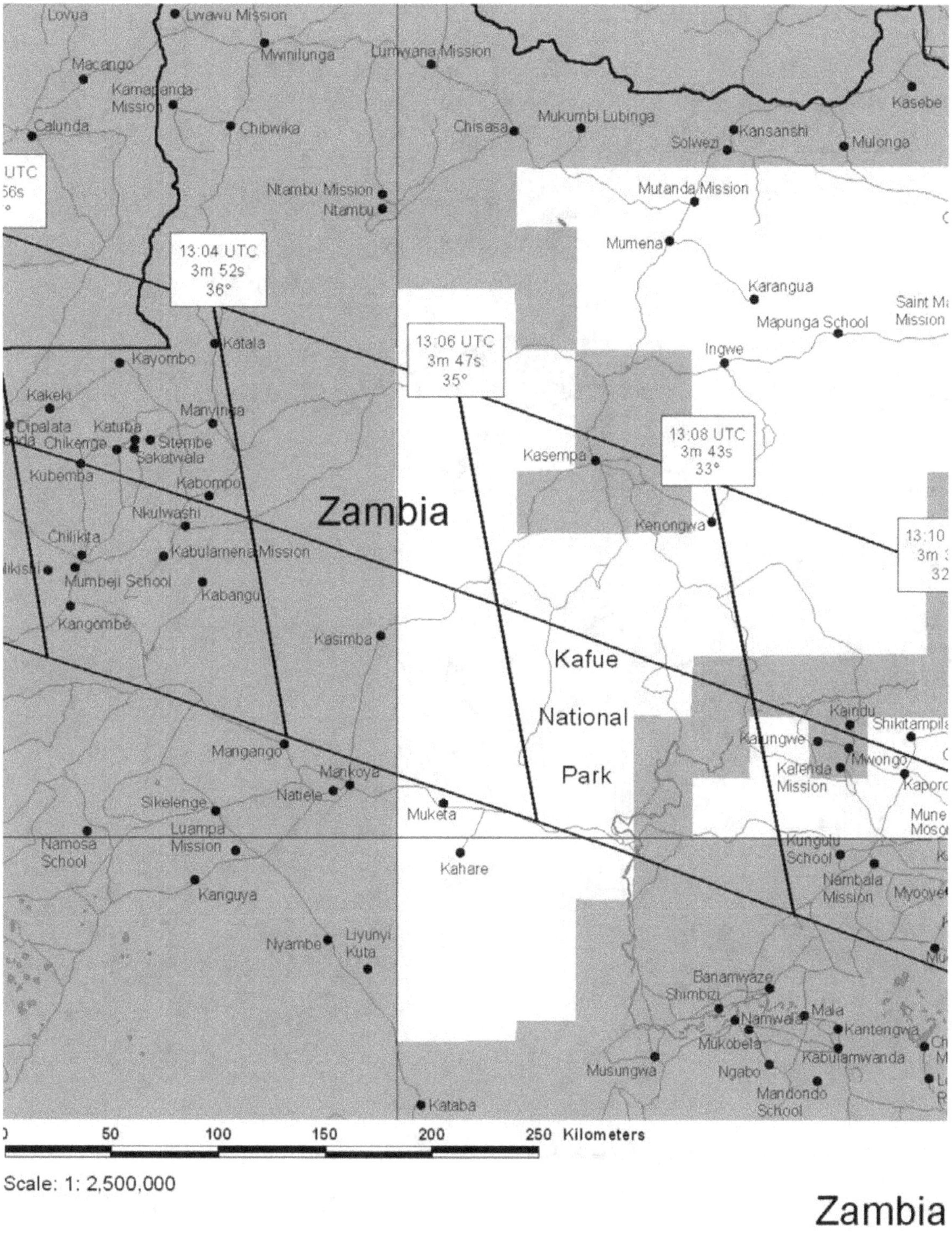

Scale: 1: 2,500,000

Zambia

Total Solar Eclipse of 2001 June 21

Map 6 - Zambia - Zimbabwe

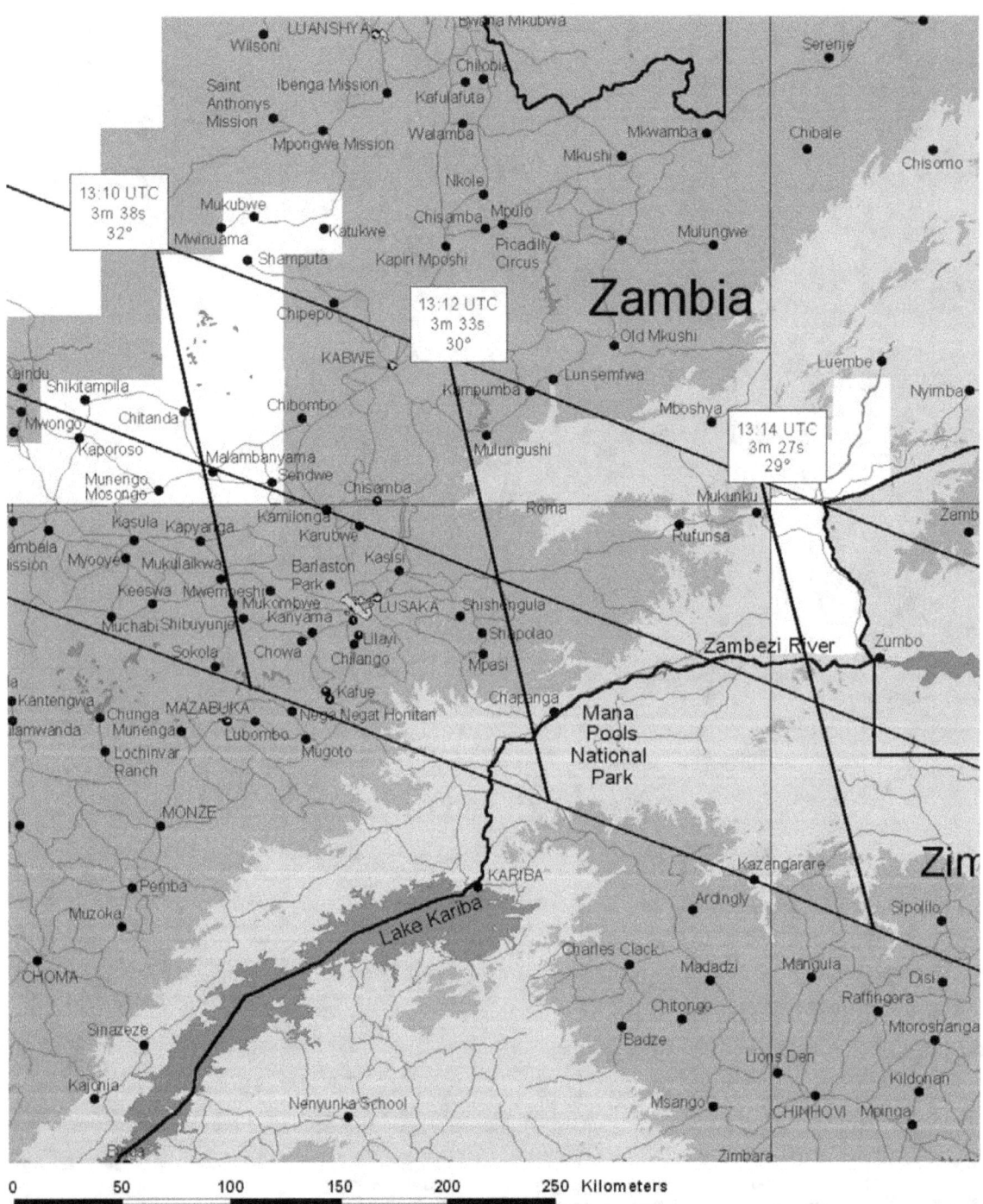

Zambia - Zimbabwe

Total Solar Eclipse of 2001 June 21

Map 7 - Zimbabwe - Mozambique

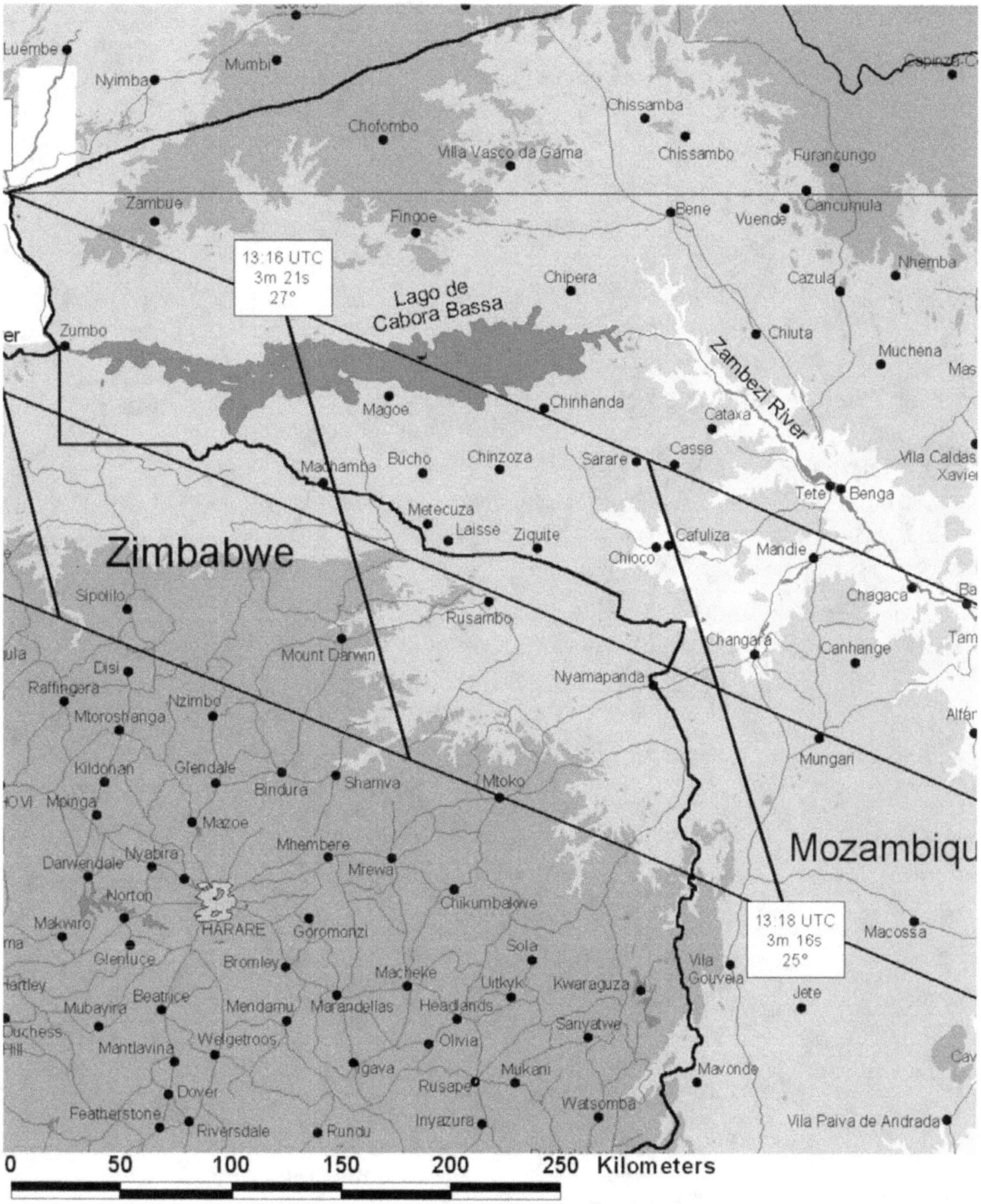

Zimbabwe-Mozambique

Total Solar Eclipse of 2001 June 21

Map 8 - Mozambique

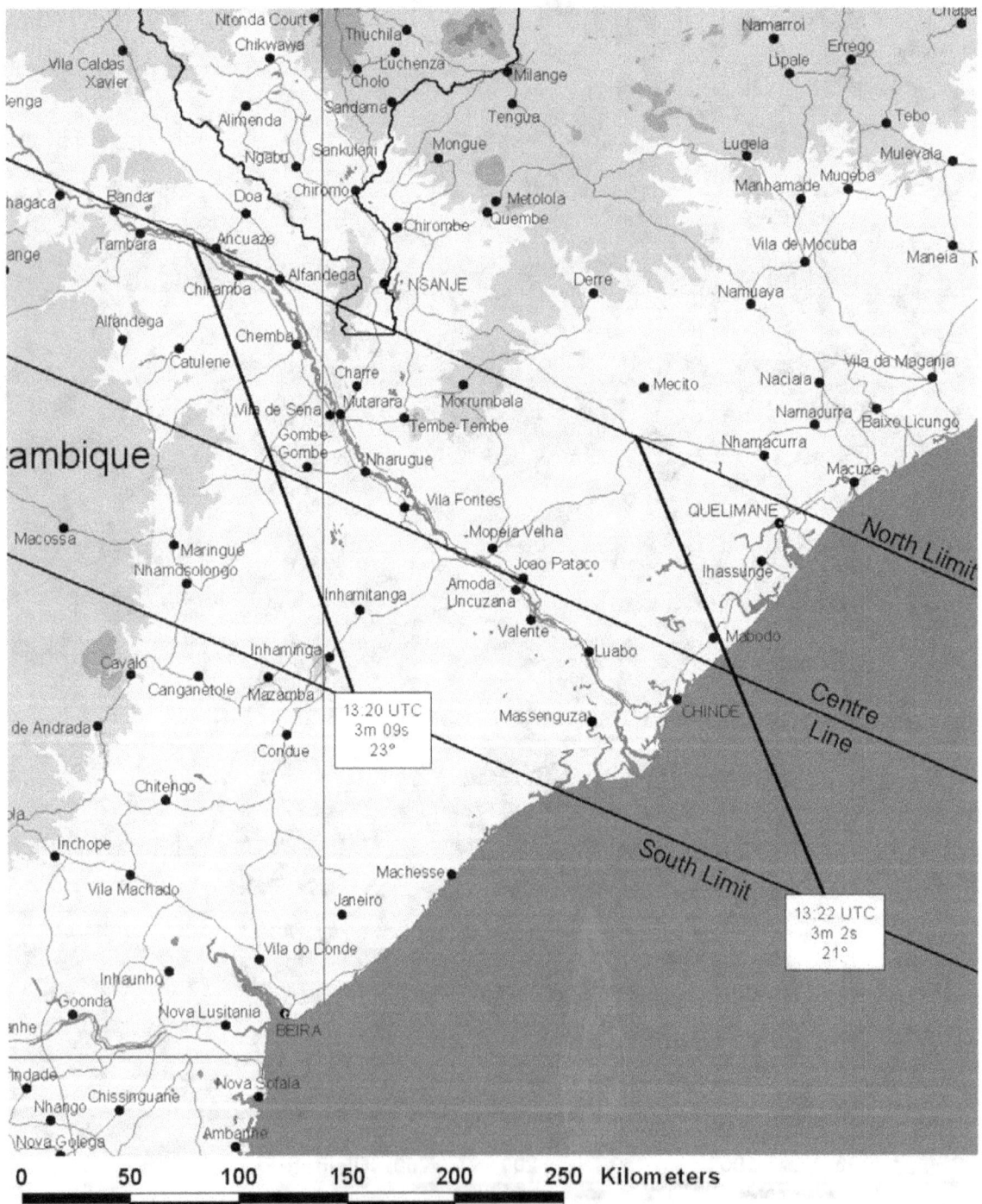

Mozambique

Total Solar Eclipse of 2001 June 21

Map 9 - Western Madagascar

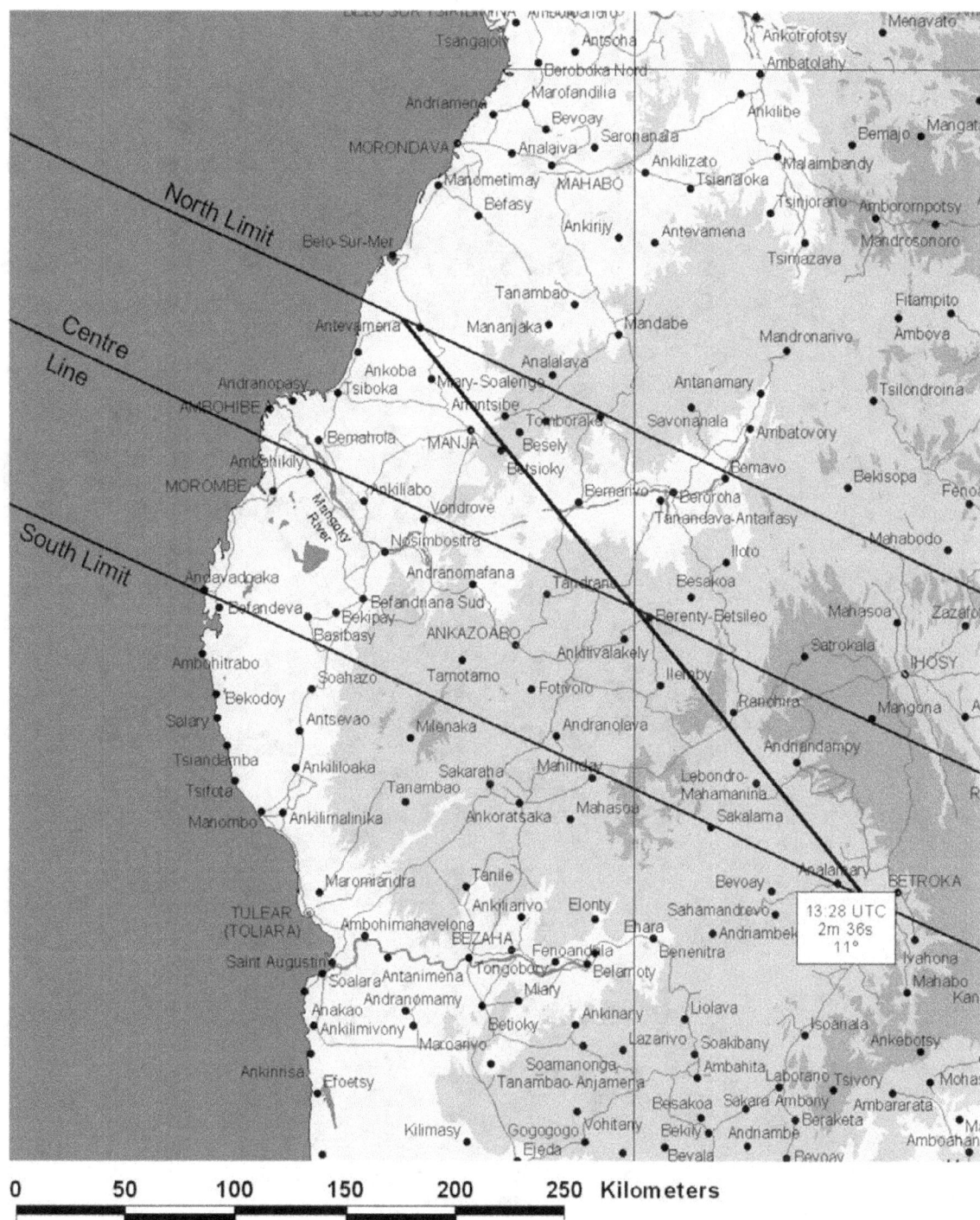

13:28 UTC
2m 36s
11°

0   50   100   150   200   250   Kilometers

Western Madagascar

Total Solar Eclipse of 2001 June 21

Map 10 - Eastern Madagascar

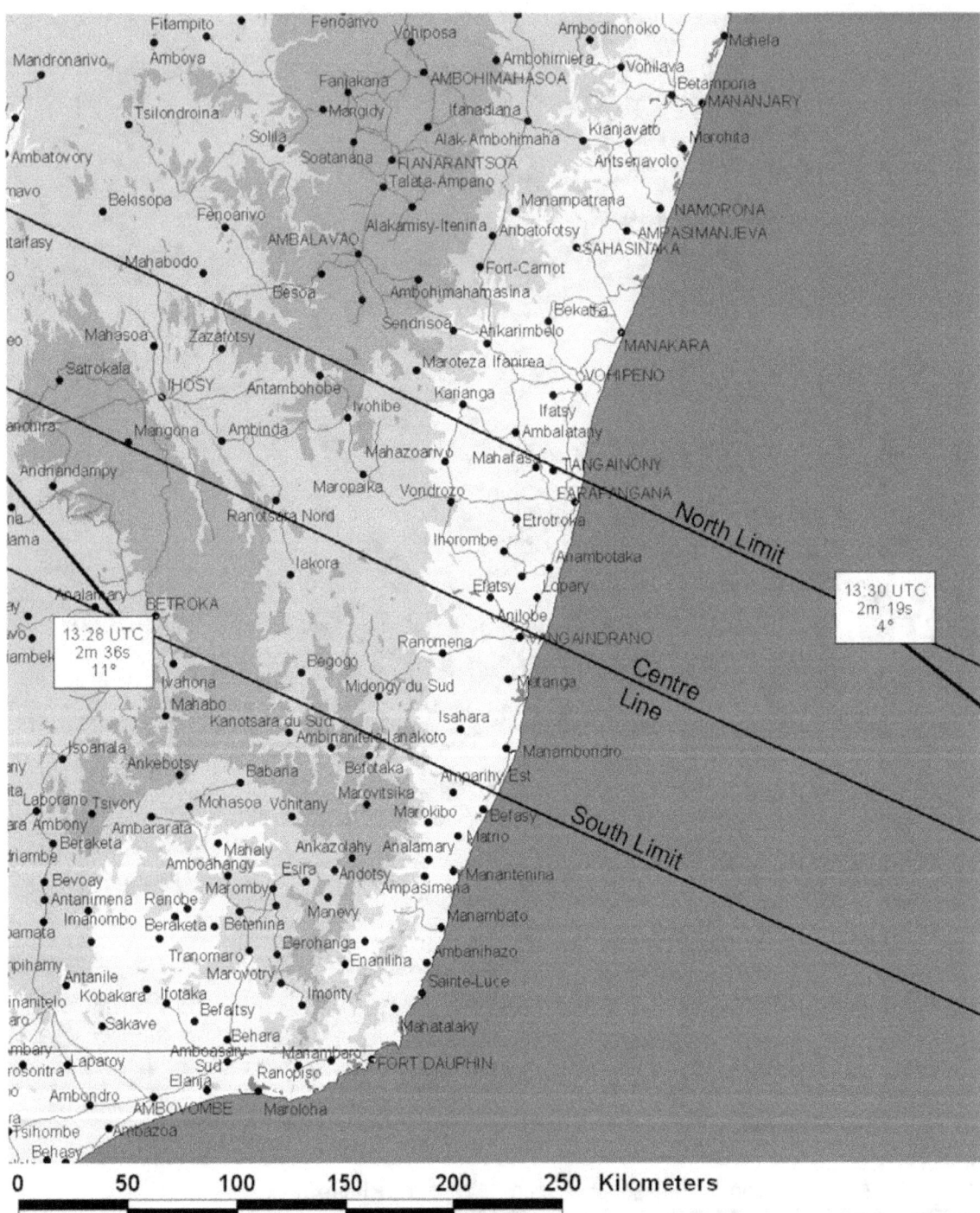

Eastern Madagascar

REQUEST FORM FOR NASA ECLIPSE BULLETINS

NASA eclipse bulletins contain detailed predictions, maps and meteorology for future central solar eclipses of interest. Published as part of NASA's Technical Publication (TP) series, the bulletins are prepared in cooperation with the Working Group on Eclipses of the International Astronomical Union and are provided as a public service to both the professional and lay communities, including educators and the media In order to allow a reasonable lead time for planning purposes, subsequent bulletins will be published 18 to 24 months before each event. Comments, suggestions and corrections are solicited to improve the content and layout in subsequent editions of this publication series.

Single copies of the bulletins are available at no cost and may be ordered by sending a 9 x 12 inch SASE (self addressed stamped envelope) with sufficient postage for each bulletin (12 oz. or 340 g). Use stamps only since cash or checks cannot be accepted. Requests within the U. S. may use the Postal Service's Priority Mail for $3.20. Please print either the eclipse date (year & month) or NASA publication number in the lower left corner of the SASE and return with this completed form to either of the authors. Requests from outside the U. S. and Canada may use ten international postal coupons to cover postage. Exceptions to the postage requirements will be made for international requests where political or economic restraints prevent the transfer of funds to other countries. Professional researchers and scientists are exempt from the SASE requirements provided the request comes on their official or institutional stationary.

Permission is freely granted to reproduce any portion of this NASA Reference Publication All uses and/or publication of this material should be accompanied by an appropriate acknowledgment of the source.

Request for: <u>NASA TP-1999-209484 — Total Solar Eclipse of 2001 June 21</u>

Name of Organization: _____

(in English, if necessary):   _____

Name of Contact Person: _____

Address: _____

_____

City/State/ZIP: _____

Country: _____

E-mail: _____

Type of organization: ___ University/College   ___ Observatory   ___ Library
(check all that apply)   ___ Planetarium   ___ Publication   ___ Media
          ___ Professional   ___ Amateur   ___ Individual

Size of Organization: _____ (Number of Members)

Activities: _____

_____

_____

* *

Return Requests and Comments to:	Fred Espenak or	Jay Anderson
	NASA/GSFC	Environment Canada
	Code 693	123 Main Street, Suite 150
	Greenbelt, MD 20771	Winnipeg, MB,
	USA	CANADA R3C 4W2

E-mail: espenak@gsfc.nasa.gov   E-mail: jander@cc.umanitoba.ca
Fax: (301) 286-0212   Fax: (204) 983-0109

1999 Oct